How Euler Did It

Leonhard Euler (1707–1783)

How Euler Did It

C. Edward Sandifer
Western Connecticut State University

Published and Distributed by
The Mathematical Association of America

© *2007 by*
The Mathematical Association of America (Incorporated)

Library of Congress Catalog Card Number 2007927658

ISBN: 978-0-88385-563-8

Printed in the United States of America

Current Printing (last digit):
10 9 8 7 6 5 4 3 2 1

SPECTRUM SERIES

The Spectrum Series of the Mathematical Association of America was so named to reflect its purpose: to publish a broad range of books including biographies, accessible expositions of old or new mathematical ideas, reprints and revisions of excellent out-of-print books, popular works, and other monographs of high interest that will appeal to a broad range of readers, including students and teachers of mathematics, mathematical amateurs, and researchers.

777 Mathematical Conversation Starters, by John de Pillis

99 Points of Intersection: Examples—Pictures—Proofs, by Hans Walser. Translated from the original German by Peter Hilton and Jean Pedersen

All the Math That's Fit to Print, by Keith Devlin

Carl Friedrich Gauss: Titan of Science, by G. Waldo Dunnington, with additional material by Jeremy Gray and Fritz-Egbert Dohse

The Changing Space of Geometry, edited by Chris Pritchard

Circles: A Mathematical View, by Dan Pedoe

Complex Numbers and Geometry, by Liang-shin Hahn

Cryptology, by Albrecht Beutelspacher

The Early Mathematics of Leonhard Euler, by C. Edward Sandifer

The Edge of the Universe: Celebrating 10 Years of Math Horizons, edited by Deanna Haunsperger and Stephen Kennedy

Five Hundred Mathematical Challenges, by Edward J. Barbeau, Murray S. Klamkin, and William O. J. Moser

The Genius of Euler: Reflections on his Life and Work, edited by William Dunham

The Golden Section, by Hans Walser. Translated from the original German by Peter Hilton, with the assistance of Jean Pedersen.

The Harmony of the World: 75 Years of Mathematics Magazine, edited by Gerald L. Alexanderson with the assistance of Peter Ross

How Euler Did It, by C. Edward Sandifer

I Want to Be a Mathematician, by Paul R. Halmos

Journey into Geometries, by Marta Sved

JULIA: a life in mathematics, by Constance Reid

The Lighter Side of Mathematics: Proceedings of the Eugène Strens Memorial Conference on Recreational Mathematics & Its History, edited by Richard K. Guy and Robert E. Woodrow

Lure of the Integers, by Joe Roberts

Magic Numbers of the Professor, by Owen O'Shea and Underwood Dudley

Magic Tricks, Card Shuffling, and Dynamic Computer Memories: The Mathematics of the Perfect Shuffle, by S. Brent Morris

Martin Gardner's Mathematical Games: The entire collection of his Scientific American columns

The Math Chat Book, by Frank Morgan

Mathematical Adventures for Students and Amateurs, edited by David Hayes and Tatiana Shubin. With the assistance of Gerald L. Alexanderson and Peter Ross

Mathematical Apocrypha, by Steven G. Krantz

Mathematical Apocrypha Redux, by Steven G. Krantz

Mathematical Carnival, by Martin Gardner

Mathematical Circles Vol I: In Mathematical Circles Quadrants I, II, III, IV, by Howard W. Eves

Mathematical Circles Vol II: Mathematical Circles Revisited and Mathematical Circles Squared, by Howard W. Eves

Mathematical Circles Vol III: Mathematical Circles Adieu and Return to Mathematical Circles, by Howard W. Eves

Mathematical Circus, by Martin Gardner

Mathematical Cranks, by Underwood Dudley

Mathematical Evolutions, edited by Abe Shenitzer and John Stillwell

Mathematical Fallacies, Flaws, and Flimflam, by Edward J. Barbeau

Mathematical Magic Show, by Martin Gardner

Mathematical Reminiscences, by Howard Eves

Mathematical Treks: From Surreal Numbers to Magic Circles, by Ivars Peterson

Mathematics: Queen and Servant of Science, by E.T. Bell

Memorabilia Mathematica, by Robert Edouard Moritz

Musings of the Masters: An Anthology of Mathematical Reflections, edited by Raymond G. Ayoub

New Mathematical Diversions, by Martin Gardner

Non-Euclidean Geometry, by H. S. M. Coxeter

Numerical Methods That Work, by Forman Acton

Numerology or What Pythagoras Wrought, by Underwood Dudley

Out of the Mouths of Mathematicians, by Rosemary Schmalz

Penrose Tiles to Trapdoor Ciphers ... and the Return of Dr. Matrix, by Martin Gardner

Polyominoes, by George Martin

Power Play, by Edward J. Barbeau

The Random Walks of George Pólya, by Gerald L. Alexanderson

Remarkable Mathematicians, from Euler to von Neumann, Ioan James

The Search for E.T. Bell, also known as John Taine, by Constance Reid

Shaping Space, edited by Marjorie Senechal and George Fleck

Sherlock Holmes in Babylon and Other Tales of Mathematical History, edited by Marlow Anderson, Victor Katz, and Robin Wilson

Student Research Projects in Calculus, by Marcus Cohen, Arthur Knoebel, Edward D. Gaughan, Douglas S. Kurtz, and David Pengelley

Symmetry, by Hans Walser. Translated from the original German by Peter Hilton, with the assistance of Jean Pedersen.

The Trisectors, by Underwood Dudley

Twenty Years Before the Blackboard, by Michael Stueben with Diane Sandford

The Words of Mathematics, by Steven Schwartzman

MAA Service Center
P.O. Box 91112
Washington, DC 20090-1112
800-331-1622   FAX 301-206-9789

This book is dedicated to my wife,
Theresa,
who reads every column, and to our children,
Philip and Victoria,
for their enthusiasm and support.

Introduction

This is a collection of the forty *How Euler Did It* columns that appeared on MAAOnline between November 2003 and February 2007. They have been lightly edited and the order has been changed to try to group together related columns.

The collection opens with the most recent column, "Euler's Greatest Hits" from February 2007. It is based on a survey of participants in an MAA Short Course on Euler from the Joint Mathematics Meetings in New Orleans in January 2007. They selected Euler's "Top Ten" mathematical results. This serves as an introduction to the breadth and importance of Euler's mathematical work.

The last column in the collection is a report from May 2005 on the previous year's meeting of The Euler Society.

In between, columns are grouped thematically. First there are six columns about geometry, including two on the Euler Polyhedral Formula, $V - E + F = 2$. The first describes Euler's discovery of the formula and the controversy over whether Descartes had also discovered it. The second gives Euler's elegant proof of the formula and discusses its subtle flaw.

Then we have six columns on number theory, beginning with Euler's first proof of Fermat's Little Theorem, the first column in the series, published in November 2003. Other columns tell how Euler discovered many pairs of amicable numbers, his work on odd perfect numbers, and his early progress towards quadratic reciprocity.

Following that, there are five columns that fit under the broad umbrella of combinatorics. Topics include Venn diagrams, chess problems, integer orthogonal matrices, partitions and permutations.

Euler was sometimes described as "Analysis Incarnate," so it is not surprising that columns about analysis form the largest group by far, a total of eighteen columns. Topics are quite varied, ranging from the foundations of calculus to the Basel problem to elliptic integrals to divergent series.

Some 40% of Euler's work dealt directly with scientific topics, such as astronomy, mechanics, fluids, ballistics, and a dozen other topics. This work is very much underrepresented among these columns; Only three of them deal with applications, one on ballistics, one on marine engineering, and one about the curious role Euler played in the discovery of America.

We close the collection with the column about The Euler Society. On the one hand, this column gives me a chance to mention some of my friends whose encouragement and support has helped to make these columns so much fun to write. On the other hand, this column serves the more serious purpose of showing how my own work on Euler depends on the work of other people. Many of those people are active in The Euler Society.

After much thought, I decided not to include a biographical sketch of Euler in this volume. Short accounts of Euler's life are easily available, and this doesn't seem to be the place for a longer treatment. I try to follow the advice of the late John Fauvel and make sure that each column includes "Content, Context and Significance." I hope that readers will find the relevant biographical material embedded in each column.

History is like any other field of mathematics, and Euler is like any other specialty, in that they require a community to thrive. Interested colleagues serve as judge, audience, sounding board and inspiration. This column is a light-hearted public face of an active and vibrant research community.

Many of the columns are directly inspired by questions from that community. That is how one of my favorite columns, "Who proved e is irrational" came to be. Three or four times a year, my university hosts a workshop on reading original sources in mathematics. We call it ARITHMOS, and Fred Rickey of the US Military Academy at West Point attends regularly. Fred was wondering why some sources credit Lambert with the first proof that e is irrational, and others credit Euler. The column was my answer to Fred's question. The columns on divergent series, on the Knight's tour, on sums and products and on cannonball curves also answer questions that people have asked. I hope that people will continue to ask such questions. I still owe Richard Askey an answer.

I am fortunate to get to spend about one day a week in the libraries at Yale University reading Euler from the original 18th century journals. A great many columns are based on things I just stumble across. "A Mystery about the Law of Cosines" is my favorite example, but the columns about orthogonal matrices, isosceles triangles, Cramer's paradox, odd perfect numbers, the *Theorema arithmeticum* and false induction all started while I was looking for something else.

The column was born at the Mathfest in Boulder, Colorado in the summer of 2004. Frank Morgan's column, "Math Chat," had just ended a four-year run on MAA Online, and that left a hole in the MAA Online Columns Page. At the same time, I had recently finished the first draft of my book, *The Early Mathematics of Leonhard Euler* (Volume 1 of this series, published in 2007) and I (foolishly) thought I'd have lots of time on my hands since I wouldn't be working so hard on the book. That assumption was comically false, but it was true that I had become familiar with a lot of Euler's work that didn't fit into the book and I was going to the library every week anyway. The mathematics I read was interesting and exciting, and that made it easy to write about. One column a month is just about the right pace.

There are a great many people whose contributions made this volume possible. First, every mathematician in America should want to write a book with the MAA just for the chance to work with Elaine Pedreira and Beverly Ruedi. They do magic with a smile. The rest of the people at the MAA are wonderful as well, especially Jerry Alexanderson and the members of the Spectrum Editorial Board for their conscientious and rigorous editing. Thanks also to Fernando Gouvêa and Carol Baxter for editing and producing MAA Online, where these columns first appeared and to Don Albers, who has played many roles in the life of this project. The director of my university library, Ralph Holibaugh and his staff have made it possible for me to use the libraries at Yale. Thanks to my wife, Theresa, and my children, Philip and Victoria, for their encouragement and support. Finally, I thank the Readers, the ones for whom we write.

Ed Sandifer
Newtown, CT
March 4, 2007

Contents

I

Euler's Greatest Hits

(February 2007)

People love Top Ten lists. Johnny Carson and Dick Clark made their careers on them. The culture of the mathematical community changes more slowly than Carson's current events or Clark's pop music did, so we have fewer reasons and opportunities to make lists.

There have been a couple of examples, though. David Wells did a survey for *Mathematical Intelligencer* in 1988 that put $e^{\pi i} + 1 = 0$ at the top of the list of the most beautiful theorems in mathematics. In 2004, *Physics World* put the same formula second in their list of greatest equations, behind the Maxwell equations. *Physics World* also put $1 + 1 = 2$ eighth on their list, praising its elegance and simplicity.

We recently had a chance to do an informal survey on Euler. Rob Bradley and I organized a Short Course at the Joint Mathematics Meetings in New Orleans in January 2007. During the course, after Ron Calinger's account of Euler's life and times, but before any specific mathematical content, we gave the participants in the Short Course a ballot listing 30 candidates as "Euler's greatest theorems." We asked for additional nominations from the floor, but there weren't any. Then we asked the participants to mark the theorems they were sure should be ranked among Euler's Top Ten, and also the ones they thought should be in the Top Three.

When we checked the ballots, we counted Top Three votes as three votes each, and the other Top Ten votes as one vote each. There were 35 ballots. The results are given below (with number of votes in parentheses). Despite the flaws in the research methodology, these people were well informed in mathematics and interested in Euler, so until somebody else does another study, we can declare this to be the

Official List of Euler's Top Ten Theorems

1. (26) Basel problem: $\zeta(2) = \displaystyle\sum_{k=1}^{\infty} \frac{1}{k^2} = \frac{\pi^2}{6}$

2. (25) Polyhedral formula: $V - E + F = 2$

3. (23) $e^{\pi i} = -1$

4. (16) Königsberg bridge problem and the Knight's tour

5. (14) Euler product formula: $\displaystyle\prod_{p \text{ prime}} \frac{1}{1 - \frac{1}{p^n}} = \sum_{k=1}^{\infty} \frac{1}{k^n}$

6. (12) Euler-Lagrange necessary condition: If a function y makes $J = \displaystyle\int_a^b f(x, y, y') \, dx$ a maximum or a minimum, then $\dfrac{\partial f}{\partial y} - \dfrac{d}{dt}\left(\dfrac{\partial f}{\partial y}\right) = 0$.

7. (11) Density of primes: $\displaystyle\sum_{p \text{ prime}} \frac{1}{p}$ diverges

8. (10) Generating functions and partition numbers

9T. (9) Euler-Fermat theorem: $a^{\varphi(n)} \equiv 1 \pmod{n}$

9T. (9) Gamma function

My own Top Ten list had seven of the same ten theorems as the audience chose, and three of the Top Five, overlaps of 70% and 60% respectively. Among the Top Five, I left off $e^{\pi i} = -1$ and Königsberg, and instead included the Euler-Fermat theorem and the Euler-Lagrange necessary condition.

I'd also left Königsberg out of my Top Ten, as well as the density of primes and the gamma function. Instead, I'd included the differential equations of fluid flow, the differential equations of the motions of solid bodies, and the Euler line theorem.

Reasonable people can disagree, and people tended to vote for the theorems that were most important in their own mathematical backgrounds. Number theorists voted for Fermat's Last Theorem in the cases $n = 3$ and $n = 4$. Robin Wilson, author of a book on the history of graph theory, [BLW] voted for the Königsberg bridges, and Fred Rickey, professor of mathematics at the US Military Academy at West Point, had voted for the theorems in Euler's *Artillerie*.

There were apparently no differential geometers in the audience, as nobody at all voted for the orthogonality of principal curvatures on a surface. Typesetters seem scarce as well, since the typographically beautiful, though not very deep differential

$$de^{e^{e^{e^x}}} = de^{e^{e^{e^x}}} e^{e^{e^x}} e^{e^x} e^x \, dx$$

got only four votes. There may have been a few 19th century geometers, but they split their votes between the nine-point circle theorem and the Euler line, which came in 15th and 23rd respectively. Mathematicians do have a common body of knowledge, so there was enough consensus that the top three theorems had a substantial lead over the rest, and all three surveys agreed that $e^{\pi i} = -1$ belongs in the top three, even if they couldn't agree whether that's the better way to write it. Some prefer $e^{\pi i} + 1 = 0$.

Even beyond $e^{\pi i} + 1 = 0$, our Official List is consistent with the lists from the *Intelligencer* and from *Physics World*. The *Intelligencer* had the following Euler and Euler-related theorems in their Top Ten:

1. $e^{\pi i} = -1$

2. $V - E + F = 2$

3. Infinitude of primes. (They meant Euclid's proof, but $\sum_{p \text{ prime}} \frac{1}{p}$ can't diverge unless there are infinitely many primes.)

5. $\zeta(2) = \frac{\pi^2}{6}$

10. If p is a prime number of the form $4n + 1$, then p can be written as the sum of two squares in exactly one way.

Physics World listed another Euler-related equation. In ninth position, they put the Principle of Least Action, $\delta S = 0$, the physical and metaphysical principle that anything that happens occurs in such a way as to maximize or minimize some quantity, such as energy, momentum, time or distance. This is related to the Euler–Lagrange necessary condition because, as we have written the formulas, the quantity J is maximized or minimized when $\delta J = 0$.

Each month, this column mostly describes the details of Euler's mathematics, finding clever, interesting and beautiful ways Euler did things. We try to put those details into the contexts of Euler's times and his other work. There are enough columns (this one is # 40 in the series) that the details combine to give a fairly extensive piece of the picture of Euler's work. Still, we've made no particular effort to give an overall picture of his work, to identify which of that work is his best, or to answer the questions "Why was Euler great?" and "Why is he famous?" Now, armed with the Official List, we can look at the columns so far and gauge how much of the Big Picture we have looked at by comparing column topics with the items on the List. There is, of course, a huge bias towards mathematics in this view, and it woefully neglects Euler's work in other fields. Only about 40% of his books and papers are mathematics. The other 60% range over mechanics, optics, astronomy, magnetism, acoustics, philosophy and half a dozen other topics. Anyway, how are we doing in the Top Ten?

1. The Basel Problem

Since William Dunham [D] gives such a beautiful account of Euler's famous infinite-product solution to the Basel Problem, we've not tried to write a column about it. In 1741, though, Euler gave a completely different solution, and that was the subject of the March 2004 column, "Basel Problem with Integrals." Also, we described Euler's numerical solution, finding the sum of the reciprocals of the squares to six decimal places, in "Estimating the Basel Problem," December 2003.

2. $V - E + F = 2$

We gave a detailed analysis of the theorem and the flaws in Euler's proof in two consecutive columns, June and July 2004, "$V - E + F = 2$," parts 1 and 2.

3. $e^{\pi i} = -1$

I have a problem with this. Though I agree that it is a beautiful and important result, I am not convinced that we are right to attribute it to Euler.

First, I've never seen Euler state the fact in this way. However, in his very first letter to Christian Goldbach in October 1729, the 22-year-old Euler wrote that a certain sum

is equal to this: $\frac{1}{2}\sqrt{\sqrt{-1} \cdot \ln(-1)}$, which is equal to the side of the square equal to the circle with diameter equal to 1.

We can decode this. A circle with diameter 1 has radius $\frac{1}{2}$, so its area is $\frac{\pi}{4}$. The square root of that is $\frac{1}{2}\sqrt{\pi}$. So, Euler is claiming that $\sqrt{\sqrt{-1} \cdot \ln(-1)} = \sqrt{\pi}$. Hitting this with a little bit of algebra gives us

$$i \ln(-1) = \pi$$
$$\ln(-1) = -\pi i$$
$$-1 = e^{-\pi i}$$

Now, take reciprocals of both sides to get

$$e^{\pi i} = -1.$$

So, Euler knew something easily equivalent to the formula as early as 1729. Euler probably would have claimed credit for the discovery if he'd figured it out for himself, so I think he learned it from Johann Bernoulli.

Moreover, it's pretty clear that Roger Cotes (1682–1716) knew the generalization, $e^{i\theta} = \cos\theta + i\sin\theta$ before Euler came on the scene.

It's a beautiful result, but until I get this attribution issue straightened out, there probably won't be a column on $e^{\pi i} = -1$.

4. Königsberg bridges and the Knight's tour

There are many descriptions of the Königsberg bridge problem. My favorite is [BLW]. With such fine expositions easily available, it doesn't make sense to devote a column to Königsberg. On the other hand, we described Euler's work on the "Knight's Tour" in April 2006.

5. Euler product formula

We described Euler's proof of the product sum formula in March 2006, "Infinitely Many Primes," though the point of that column was more closely related to item 7 on our Top Ten list, the density of primes. We also mentioned it in the column "Formal Sums and Products," July 2006.

6. Euler-Lagrange necessary condition

The calculus of variations hasn't appeared yet in any of our columns. It is a beautiful and deep subject, but it has a reputation for being difficult for non-specialists to understand and appreciate. Perhaps we can find an opportunity in a future column.

7. Density of primes

The column from March 2006, "Infinitely Many Primes," describes most of Euler's work on this result in some detail.

8. Generating functions and the partition problem

We've written about this twice, first in "Roots by Recursion," June 2005, and then some different aspects of the same topic in "Philip Naudé's Problem," October 2005.

9T. The Euler-Fermat theorem

Back in November 2003, we devoted our very first column to Euler's proof of "Fermat's Little Theorem." We haven't returned to the topic, though, so we haven't described either the Euler phi-function or the Euler-Fermat theorem.

9T. Gamma function

We haven't written a column about the gamma function, either. It would make a good column.

Summary

Four of the theorems, Basel problem, $e^{\pi i} = -1$, Euler-Lagrange necessary condition and gamma function, have not been central topics of columns yet.

Two of the theorems, Königsberg/Knight's tour and Euler-Fermat theorem, have been about half covered.

Four of them, $V - E + F = 2$, Euler product formula, density of primes and generating functions/partitions have been well covered.

It looks like a lot of the picture is still missing. It will be a while before we run out of material for more columns.

References

[D] Dunham, William, *Euler: The Master of Us All*, MAA, Washington, DC, 1999.

[BLW] Biggs, Norman, Keith Lloyd and Robin Wilson, *Graph Theory 1736–1936*, Oxford University Press, 1986.

Part I

Geometry

2

V, E and F, Part I

(June 2004)

About 15 years ago, the *Mathematical Intelligencer* polled its readers to choose the ten most beautiful theorems in mathematics. The top two were results of Euler. The "Euler Identity," $e^{\pi i} = -1$ ranked at the top of the list, with the "Euler formula" $V - E + F = 2$ right below it. This month's and next month's columns, "V, E and F, Parts 1 and 2" are about the Euler formula.

Euler's formula tells us that if we have a polyhedron that satisfies certain conditions, and if the polyhedron has V vertices, E edges and F faces, then $V - E + F = 2$.

This is, indeed, a beautiful and popular result. Elementary school teachers use the result to try to lead their students to discover mathematical truths, and professional mathematicians use its generalizations in their research in fields such as algebraic topology and differential geometry.

Unfortunately, those "certain conditions" we glossed over can get a little bit nasty. We have to exclude, for example, polyhedra with holes in them, like donuts, and polyhedra with disconnected interiors, like two cubes joined at a vertex. It can get discouraging, as Imre Lakatos showed so brilliantly in his 1976 book *Proofs and Refutations* [L]. After reading that book and seeing so many challenging counterexamples, we are tempted to despair and say that the formula is true just for those polyhedra for which it is true.

Euler could not have known most of the difficult examples that Lakatos uses. Since most people have not read his original papers from 1750 and 1751, written in Latin, but many have read Lakatos and other modern sources, some half-truths have arisen about what Euler did and what he proved in those two papers. Among those half-truths are

1. Euler got it wrong, because he thought his formula applies to *all* polyhedra.
2. Euler couldn't provide a proof for his formula.
3. Euler gave a proof, but the proof was wrong.
4. It shouldn't be Euler's Formula at all, since Descartes did it first.

In fact, half of these statements are more or less half true. Our purpose here is to describe what Euler did and how he did it, and to try to figure out which parts of each of these "half-truths" are true.

Euler in the late 1740s and early 1750s enjoyed some of the richest creative years of any mathematician or scientist ever. Almost single-handedly he filled the pages of two of the world's most important scientific journals, the *Mémoires* of the Berlin Academy and the *Novi Commentarii* of the St. Petersburg Academy. In 1750, for example, he published 35 papers, and another 20 in 1751. In 1750, he turned some of his attention to the properties of solids, a subject he called "stereometry." The result was his first paper on the subject, number 230 on the Eneström index[1] [E230], titled "Elementa doctrinae solidorum," or "Elements of the doctrine of solids." A year later he wrote a shorter sequel [E231], "Demonstratio nonnullarum insignium proprietatum, quibus solida hedris planis inclusa sunt praedita," or "Proof of some of the properties of solid bodies enclosed by planes." These two papers were published back-to-back in the 1752/53 volume of the *Novi Commentarii* of the St. Petersburg Academy, which, subject to the typical publication delay of the times, appeared in print in 1758.

E230 — Elements of the doctrine of solids

Euler begins E230 with an eloquent description of a grand plan to put the geometry of solids on the same elegant foundations as Euclid did for plane geometry. We get the impression that he hoped to write a great number of papers on the new subject of stereometry, though, in fact, he only wrote these two.

Then, by analogy with polygons, which consist of points and lines, he tells us that the solids he wants to study consist of points, lines and planes. The points are each solid angles, formed where three or more planes come together. He calls them *anguli solidi* and denotes their number by S. We will use the English language tradition and call them vertices, and denote them by V. The faces he calls *hedra*, denoted by H. We will use F.

The lines are a problem for Euler, though. He tells us that they do not have a proper name, so he decides to call them *acies*, which translates as keenness, edge, penetration, insight, or battle line. What we now call an edge hadn't been named yet. Where he used A to count them, we will use E.

Most solids don't have names. Euler invests some time in developing a nomenclature, that didn't catch on, based on the numbers of vertices and faces. For example, a triangular prism, with six vertices and five sides, he called a *pentaedrum hexagonum*, or "five-faced hexagon."

Eventually, seven pages into the 30-page paper, Euler gets to some theorems. He wants to count the number of "plane angles" in a solid. It is not a quantity we still use very often, and Euler never assigns the number a variable name, so let's call it P. For example, a cube has six 4-sided faces, making a total of 24 plane angles, so for a cube, $P = 6 \cdot 4 = 24$. Now, Euler's Proposition 1 is

> *Propositio 1: In quovis solido numerus omnium acierum est semissis numeri omnium angulorum planorum, qui in cunctis hedris ambitum eius constituentibus reperiuntur.*

---

[1] In the early 1900s, Gustav Eneström prepared an authoritative list of Euler's 866 publications, in order of publication date. Whenever we refer to E-numbers or Eneström numbers, we are referring to that list. The list is available online at The Euler Archive.

Proposition 1: In any solid, the number of edges is half of the number of plane angles which are located in the corners of the faces.

As a formula, Euler is telling us that $E = P/2$. Euler adds corollaries to this proposition. In the days before subscripts, he let a be the number of triangular faces on the solid, b the number of quadrilaterals, c the number of pentagonal faces, etc. Then we have two more equations:

$$F = a + b + c + d + e + \text{etc.}$$
$$P = \frac{3a + 4b + 5c + 6d + 7e + \text{etc}}{2}.$$

Euler continues, proving two propositions that we usually prove as corollaries of the Euler formula itself, that $2E \geq 3F$ and $2E \geq 3V$.

Now Euler gets to his first big theorem:

Propositio 4: In omni solido hedris planis incluso aggretatum ex numero angulorum solidorum et ex numero hedrarum binario excedit numerum acierum

Proposition 4: In any solid enclosed by planes, the sum of the number of solid angles and the number of faces exceeds the number of edges by 2.

As a formula, this is $V + F = E + 2$. Euler never writes it as $V - E + F = 2$.

Then Euler seems to begin a proof of the theorem! But the proof begins with an apology: "I have not been able to find a firm proof of this theorem." Instead, he works through a series of progressively more complicated and general examples, and ends with a check of the five Platonic solids. But he admits that, however convincing these examples may be, they do not make a proof.

Assuming that the proposition is true, Euler does prove a result that is essential to studies of the Four Color Theorem:

Proposition 7: There cannot exist a solid all of whose faces have six or more sides, nor can there exist a solid all of whose solid angles are formed by six or more plane angles.

Finally, he turns to a theorem that he regards as important as $V - E + F = 2$, but which is practically unknown today. He gives two versions of the same result:

Proposition 8: The sum of all the plane angles which are in a given solid is equal to four right angles for each unit by which the number of edges exceeds the number of solid angles.

Proposition 9: The sum of all the plane angles which occur on the outside of a given solid is equal to eight less than four right angles for each solid angle.

If we let S be the sum of all the plane angles, measured as multiples of 90°, then Proposition 8 tells us that $S = 4E - 4V$ right angles, and Proposition 9 says that $S = 4V - 8$ right angles. Euler gives a correct proof of Proposition 8, then uses the Euler formula to prove Proposition 9 as a consequence.

This brings us to the end of E230, Euler's first paper, written in 1750. We can pause briefly to see how Euler is doing with regard to the four "half-truths" listed above

1. Euler got it wrong, because he thought his formula applies to *all* polyhedra.
   This still seems half-true. Sometimes, as in Proposition 1, Euler claims his theorem
   is true for *any* solid. When he does that, he sometimes over-reaches. However, for his
   ground-breaking claims, as in Proposition 4, that $V + F = E + 2$, Euler specifically
   states the condition that the polyhedron be *enclosed by planes*. If we interpret this as
   implying that the polyhedron is convex, then his claims are true.

2. Euler couldn't provide a proof for his formula.
   Indeed, in E230, Euler says that he cannot give a good proof. Let's wait for E231,
   though.

3. Euler gave a proof, but the proof was wrong.
   Let's wait for E231 for this, too.

4. It shouldn't be Euler's Formula at all, since Descartes did it first.
   More on this later. It is a remarkable story.

This story is getting long. Rather than burden the reader with a double-length column,
let's think about this for a month before we turn to E231, Euler's second paper on
polyhedra.

References

[C]   Cromwell, Peter R., *Polyhedra*, Cambridge University Press, 1997.

[D]   Descartes, René, *Progymnasmata de solidorum elementis*, in *Œuvres de Descartes*, vol. X,
         pp. 265–276, Adam and Tannery, reprinted by Vrin, Paris, 1996.

[E230] Euler, Leonhard, Elementa doctrinae solidorum, *Novi commentarii academiae scientiarum
         Petropolitanae* 4 (1752/3) 1758, pp. 109–140, reprinted in *Opera Omnia* Series I vol. 26
         pp. 71–93.

[E231] ——, Demonstratio nonnullarum insignum proprietatum quibus solida hedris planis inclusa
         sunt praedita, *Novi commentarii academiae scientiarum Petropolitanae* 4 (1752/3) 1758,
         pp. 140–160, reprinted in *Opera Omnia* Series I vol. 26 pp. 94–108.

[L]   Lakatos, Imre, *Proofs and Refutations: The Logic of Mathematical Discovery*, Cambridge
         University Press, 1976.

3

V, E and F, Part 2

(July 2004)

Last month, we began to examine Euler's two papers on the general properties of poly-hedra, often cited as the pioneering work on the subject and one of the first contributions to the field of topology. Euler had written what is usually regarded as the first topology paper in 1736, when he wrote of the bridges of Königsburg.

Euler wrote his first paper on this subject, "Elements of the doctrine of solids," in 1750. In that paper, he claimed two main results:

Proposition 4: In any solid enclosed by planes, the sum of the number of solid angles and the number of faces exceeds the number of edges by 2.

Proposition 9: The sum of all the plane angles which occur on the outside of a given solid is equal to eight less than four right angles for each solid angle.

In E230, Euler seems to think that these two results are of about equal importance. He gives a proof of the second of these, using the first one, but he says that he is unable to find a proof of Proposition 4, the familiar Euler formula, that $V - E + F = 2$.

Last month, we listed four "half-truths" about Euler and the Euler formula, and we set about to find what truth there was in these half-truths. When we finished reading E230, we left it like this:

1. Euler got it wrong, because he thought his formula applies to *all* polyhedra.
   This still seems half-true. Sometimes, as in Proposition 1, Euler claims his theorem is true for *any* solid. When he does that, he sometimes over-reaches. However, for his ground-breaking claims, as in Proposition 4, that $V + F = E + 2$, Euler specifically states the condition that the polyhedron be *enclosed by planes*, a condition that is sufficient to make his claims true.

2. Euler couldn't provide a proof for his formula.
   Indeed, in E230, Euler says that he cannot give a good proof. Let's wait for E231, though.

3. Euler gave a proof, but the proof was wrong.
   Let's wait for E231 for this, too.

4. It shouldn't be Euler's Formula at all, since Descartes did it first.
   More on this later. It is a remarkable story.

Now, let us turn to the sequel, and see how these half-truths stand up to the rest of the story.

E231 — Proof of some of the properties of solid bodies enclosed by planes

Euler tells us right in the title of E231 that he has a proof of his formula, and that he is not considering general solid bodies, like tori or spheres, but only those "enclosed by planes." This condition is sufficient to make $V + F = E + 2$ true.

Euler opens the paper by reviewing what he considers to be the two main results of E230, that $V + F = E + 2$ and the one about the sums of the plane angles in a polyhedron. Then he begins preparations for the proof his title promises. He plans to prove $V + F = E + 2$ by a kind of reduction argument. Today we would structure the proof as a mathematical induction, but that was not yet a popular style of proof in 1751. To prepare his readers for his proof technique, he first demonstrates how it works in two dimensions, to show that the angles in a polygon with A angles sum to $2A-4$ right angles. Suppose, for example, $ABCDEFG$ is the polygon in Euler's Figure 1, implicitly assumed to be convex. It is described by its vertices. Euler dissects the polygon into triangles by drawing diagonals GB, BF, FC, CE.

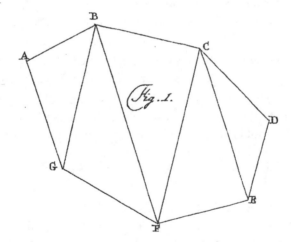

Euler goes into some detail to say that removing triangle CDE leaves a polygon with $A-1$ angles, and it decreases its angle sum by two right angles. Eventually, after n steps, for some n, we get $A - n = 3$ and are left only with triangle ABG, for which the angle sum is two right angles. Therefore, the number of right angles in the original polygon was $2n + 2$, or, equivalently, $2A - 4$.

There are easier ways to do this, but Euler wanted to prepare us for his technique in three dimensions.

Next, Euler looks for a three-dimensional analogy to the triangulation he used in two dimensions. He proposes choosing any point on the interior of the polyhedron, and

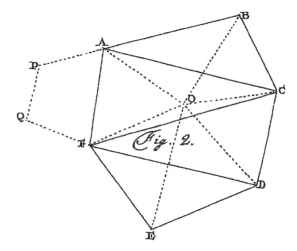

extending edges to each vertex of the polyhedron. This leads to a decomposition of the polyhedron into pyramids, with the faces of the original polyhedron as the bases of the pyramids. He tells us that this technique does not work, but he does not tell us why he thinks that. I think a modern mathematician could make it work in a way that Euler would find convincing

Thinking that this approach wouldn't work, Euler turned to a different kind of reduction plan. He decided to remove vertices. He makes some mistakes, and he makes them all in his lemma:

Proposition 1: Given any solid enclosed by planes, then a given solid angle being cut off, in the solid that remains, the number of solid angles will be one less.

His plan, illustrated in his Figure 2, is to remove the vertex labeled O. Vertex O is connected to vertices A, B, C, D, E and F, so when he removes O, he also takes a number of triangular pyramids, $OABC$, $OACF$, $OCDF$, and $ODEF$. (Notice that points P and Q shown in Figure 2 are not used in Proposition 1. Euler recycles Figure 2 in Proposition 2, where he does use them.) He then shows that the operation of removing vertex O along with all of these triangular pyramids preserves the relationship among V, E and F. He concludes that if he keeps removing vertices like this, eventually he'll have only four vertices left, that is, he'll have a triangular pyramid. Since $V + F = E + 2$ for triangular pyramids, he concludes that $V + F = E + 2$ in the original polyhedron as well.

Euler argues like this. If we remove vertex O, then V decreases by 1. Suppose that k is the number of faces that meet at O. Then k is also the number of edges that meet at O, as well as the number of sides to the (perhaps non-planar) polygon represented by $ABCDEF$ in Figure 2. Thus, the polygon triangulates to give $k - 2$ polygons, and that triangulation requires the introduction of $k - 3$ edges. Hence, removing O takes away one vertex (O), k edges (OA, OB, etc.) and k faces (OAB, OBC, OCD, etc.). However, it also adds the $k - 2$ faces of the triangulation and the $k - 3$ edges that went into the triangulation. So, in the new polyhedron, V is replaced with $V - 1$, E is replaced by $E - k + (k - 3) = E - 3$, and F is replaced by $F - k + (k - 2) = F - 2$. Thus $V - E + F$ is replaced by $V - 1 - (E - 3) + F - 2 = V - E + F$, that is, $V - E + F$ does not

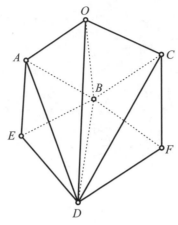

Figure 3.

change. So, if $V - E + F = 2$ after all the reductions, that is, for a triangular pyramid, then $V - E + F = 2$ for the original polygon.

That would be a beautiful proof, were it not for a subtle but serious flaw. An example is illustrated in Figure 3. If the initial polyhedron is "bounded by planes," then the polyhedron resulting after removing the vertex O is not necessarily still bounded by planes. So, we might not be able to repeat the process of removing vertices. In his Figure 3, for example, removing the vertex O according to Euler's recipe leaves two tetrahedra, $AEBD$ and $CFBD$, joined along the common edge BD. Euler would not accept this object as a polyhedron.

Polyhedra that result from removing a vertex can have other unexpected properties that can make repairing this proof more difficult than we might expect, and, even if the proof can be repaired, the fact remains that the proof given by Euler is flawed and incomplete. Still, the theorem is true.

The End of the Story

Let's review our list of half-truths.

1. Euler got it wrong, because he thought his formula applies to *all* polyhedra.
   Not true. Euler may have *thought* it applied to all polyheda, but he only *claimed* that it applied to "polyhedra bounded by planes," that is, convex polyhedra, and it does apply to them.

2. Euler couldn't provide a proof for his formula.
   Half true. Euler couldn't give a proof in his first paper, E230, and he said so, but a year later, in E231, he gives a convincing proof.

3. Euler gave a proof, but the proof was wrong.
   True. The proof Euler gives in E231, though convincing, is incorrect.

4. It shouldn't be Euler's Formula at all, since Descartes did it first.
   We think this is half-true, but it contains a grain of truth. To know why, we have to tell a story. We rely mostly on Crowell [C, pp. 181–189] for our facts.

In 1649, the year before he died, Descartes went to Sweden to tutor Princess Christina in philosophy. When he died, his possessions were sent back to France, but, just as they arrived in Paris, the box of his manuscripts fell into the river. They were mostly rescued, and dried, and some were available to scholars at the time, including Leibniz. Leibniz hand-copied several of the manuscripts, including one of sixteen pages titled *Progymnasmata de solidorum elementis*, [D] or "Exercises in the elements of solids." The original subsequently vanished, and Leibniz' copy disappeared into his papers until it was rediscovered in 1860. Thus, for over 200 years, only Leibniz knew of Descartes' work on polyhedra, and he doesn't seem to have told anybody. Nobody in Euler's time would have known of it.

The *Progymnasmata* is the first known study of general polyhedra. Even though it is not well known, it is the first. That distinction is often credited to Euler.

So, what did Descartes do? He studied something closely related to Euler's formula for the sum of the plane angles of a polyhedron. In Descartes' time, people had a concept of a solid angle called the *deficiency*. The *deficiency* of a solid angle is the amount by which the sum of the plane angles at the solid angle fall short of four right angles. In the case, for example, of a solid right angle, formed by three right angles, the deficiency will be one right angle. For a cube, which contains eight solid right angles, the total deficiency is eight right angles. Descartes' main result is that this always happens:

Theorem. *The sum of the deficiencies of the solid angles of a polyhedron is always eight right angles.*

It is an almost trivial step from this to Euler's theorem, that the sum of the plane angles is four times the number of solid angles, less eight right angles, that is $4V - 8$ right angles.

Descartes' other interesting result is more subtly related, but still remotely equivalent to $V - E + F = 2$. Descartes writes:

Dato aggregato ex omnibus angulis planis et numero facierum, numerum angulorum planorum invenire: Ducatur numerus facierum per 4, et productum addatur aggregato ex omnibus angulis planis, et totius media pars erit numeris angulorum planorum.

Given the sum of all the plane angles and the number of faces, to find the number of plane angles: The number of faces is multiplied by 4, and to the product is added the sum of all the plane angles, and the half part of this total will be the number of plane angles.

It is easy, but not obvious, to transform this rule into Euler's $V - E + F = 2$, as follows.

Let Σ denote the sum of all the plane angles in the solid, and let V be the number of vertices, what Descartes and Euler both call "solid angles." Then Descartes has just told us that $4V - \Sigma = 8$ right angles. Further, let F be the number of faces and P the number of plane angles.

It is easy for us to see that the number of plane angles is twice the number of edges, that is, $P = 2E$.

Now, Descartes tells us that given Σ and F, we are to find P. He uses words amounting to the formula

$$\frac{4F + \Sigma}{2} = P.$$

Substituting $P = 2E$ and $\Sigma = 4V - 8$, we get

$$\frac{4F + 4V - 8}{2} = 2E,$$

which is easy to transform into

$$V - E + F = 2.$$

So, did Descartes find the Euler formula? It would make a good debate.

References

[C]   Cromwell, Peter R., *Polyhedra*, Cambridge University Press, 1997.

[D]   Descartes, René, *Progymnasmata de solidorum elementis*, in *Œuvres de Descartes*, vol. X, pp. 265–276, Adam and Tannery, reprinted by Vrin, Paris, 1996.

[E230] Euler, Leonhard, Elementa doctrinae solidorum, *Novi commentarii academiae scientiarum Petropolitanae* 4 (1752/3) 1758, pp. 109–140, reprinted in *Opera Omnia* Series I vol. 26 pp. 71–93.

[E231] ——, Demonstratio nonnullarum insignum proprietatum quibus solida hedris planis inclusa sunt praedita, *Novi commentarii academiae scientiarum Petropolitanae* 4 (1752/3) 1758, pp. 140–160, reprinted in *Opera Omnia* Series I vol. 26 pp. 94–108.

[J]   Jonquières, E. de, Ecrit posthume de Descartes intitulé de solidorum elementis, *Bibliotheca Mathematica, Neue Folge*, 4 (1890), pp. 43–55.

[L]   Lakatos, Imre, *Proofs and Refutations: The Logic of Mathematical Discovery*, Cambridge University Press, 1976.

Figures were downloaded from wwwEulerArchive.org.

4

19th Century Triangle Geometry

(May 2006)

When we read about Euler, or about any other historical figure, we must remember that he lived in his own times. The 18th century was very different from the 21st in ways that we hardly ever think about. There are the obvious differences; now we have the internet, iPods, airplanes and automobiles. I am fond of reminding my students that we also have indoor plumbing, grocery stores and paper money. So, when we read Euler, we must try to understand how the problems he works on and the techniques he uses are embedded in his own times, and not in ours. He was speaking to and writing for an 18th century audience and we are lucky that the things he was saying are still useful and interesting today. So, when we find Euler seeming to use 17th century techniques to solve a 19th century problem, we might raise an eyebrow.

Euler's paper *Geometrica et sphaerica quaedam*, [E749] which translates uninformatively as "Certain geometric and spheric things," is such a paper. Its main result is a theorem in triangle geometry, a subject that was extremely popular and important in the late 19th century. The main results in triangle geometry are summarized in excellent books like that of Coxeter and Greitzer [CG].

In contrast, one of the major mathematical themes of Euler's era was the gradual evolution, led by Euler himself, from a mathematics based on techniques and objects of geometry to one based on algebra and analysis. Though this is not the place to dwell too much on this point, we note that at the beginning of the 18th century, mathematicians called themselves "geometers" and they used calculus to study curves, in the style of L'Hôpital. By the end of the century, they called themselves "mathematicians" and "analysts" and used calculus to study functions. In E749, Euler gives three proofs of a 19th century result, but his third proof, clearly his favorite of the three, is a proof with a 17th century flavor.

Euler apparently wrote E749 in 1780. Euler's son-in-law Nicolas Fuss, presented it to the Academy in St. Petersburg, along with three other papers, on May 1 of that year. In 1780, Euler was 73 years old and he no longer attended the meetings of the Academy himself. Euler's last meeting seems to have been on January 16, 1777, after which Euler sent his papers in to the Academy with his assistants. In 1780, Euler had been blind

Leonhard Euler

for almost 15 years, and he had a team of assistants to whom he dictated hundreds of manuscripts. One of the portraits of Euler, shown above, has a sub-portrait, a smaller rectangle beneath the oval of the main portrait. The sub-portrait shows two men, one with pen and paper, sitting at a table. Apparently it pictures Euler dictating to one of his assistants, probably his son, Johann Albrecht, because Euler himself could no longer read or write.

Let us turn to the mathematics. Euler gives us the triangle $\triangle ABC$ shown in Figure 1, cut by concurrent segments Aa, Bb and Cc, where points given by lower case letters are on the sides opposite the vertices given by the upper case letters. Note that the point where the segments intersect is named O. Euler asks, given the lengths of the segments AO, Oa, BO, Ob, CO and Oc, can he reconstruct the triangle?

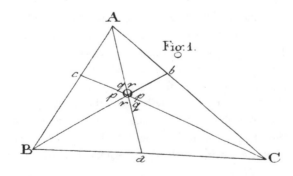

He finds that there will not be such a triangle unless certain conditions on the ratios of the lengths of the segments are satisfied, and gives us the following:

Theorem. *If in any triangle ABC are drawn from each angle to the opposite side any straight lines Aa, Bb, Cc cutting each other at a common point O, then they will always satisfy this property, that*

$$\frac{AO}{Oa} \cdot \frac{BO}{Ob} \cdot \frac{CO}{Oc} = \frac{AO}{Oa} + \frac{BO}{Ob} + \frac{CO}{Oc} + 2. \tag{1}$$

Euler's proof is rather long and not very elegant. We debated omitting it but eventually decided to include it so that later we could admire how much more elegant his third proof is.

Proof. Take $AO = A$, $BO = B$, $CO = C$, $Oa = a$, $Ob = b$, $Oc = c$ and label the six angles around O as shown. Note that $p + q + r = 180°$. As was typical in his times, Euler expects us to be able to use context to distinguish the points A, B, C, a, b and c from the lengths with the same names.

We can find the area of $\triangle AOc$ to be $\triangle AOc = \frac{1}{2} Ac \sin q$. Similarly, $\triangle BOc = \frac{1}{2} Bc \sin p$ and $\triangle AOB = \frac{1}{2} AB \sin(p + q)$.

Since $\sin(p + q) = \sin r$, and the areas of the first two triangles sum to the third, we get:

$$AB \sin r = Ac \sin q + Bc \sin p.$$

Similarly, for the other two pairs of triangles, we get

$$BC \sin p = Ba \sin r + Ca \sin q$$

$$CA \sin q = Cb \sin p + Ab \sin r.$$

Dividing these equations by ABc, aBC and AbC respectively gives

$$\frac{\sin r}{c} = \frac{\sin q}{B} + \frac{\sin p}{A}$$
$$\frac{\sin p}{a} = \frac{\sin r}{C} + \frac{\sin q}{B} \tag{2}$$
$$\frac{\sin q}{b} = \frac{\sin p}{A} + \frac{\sin r}{C}.$$

Euler pauses to point out the pattern in these three equations.

Define α, β and γ by the equations $A = \alpha a$, $B = \beta b$, $C = \gamma c$ and then define P, Q, R by the equations

$$P = \frac{\sin p}{A} = \frac{\sin p}{\alpha a}$$
$$Q = \frac{\sin q}{B} = \frac{\sin q}{\beta b} \tag{3}$$
$$R = \frac{\sin r}{C} = \frac{\sin r}{\gamma c}.$$

Then the three formulas in (2) transform into

$$\gamma R = P + Q, \quad \alpha P = Q + R, \quad \beta Q = R + P$$

This gives us the ratios

$$P : R = \gamma + 1 : \alpha + 1$$
$$Q : P = \alpha + 1 : \beta + 1$$
$$R : P = \beta + 1 : \gamma + 1.$$

This has a nice pattern, too. From this we get the triple proportion

$$P : Q : R = \frac{1}{\alpha + 1} : \frac{1}{\beta + 1} : \frac{1}{\gamma + 1}. \tag{4}$$

though, to the modern eye, that doesn't look like a very convenient way to write anything.

Now, the first of our three equations in (3) gives

$$R = \frac{P + Q}{\gamma}.$$

From the second equation we get $R = \alpha P - Q$. Put these two values into the ratio between P and Q given in (4) and we get

$$\frac{P}{Q} = \frac{\gamma + 1}{\alpha \gamma - 1}.$$

Since also $\frac{P}{Q} = \frac{\beta+1}{\alpha+1}$, all this multiplies out to give

$$\alpha \beta \gamma = \alpha + \beta + \gamma + 2.$$

Substituting back the triangle measurements for the Greek letters gives the result of the theorem.   Q.E.D.

Here is an example of how we might be misled by reading an old theorem with modern eyes. Now we think of this theorem and the corollaries that we will see below as properties of triangles. That's not what Euler had in mind, though. This theorem gives a necessary property that the lengths of the six given segments must satisfy in order for him to solve the problem of finding the triangle that gives rise to those six lengths. That is to say, he still wants to solve the following:

Problem.   Given that the parts of a triangle are lengths A, B, C, a, b, c as described above, to construct the triangle.

Euler's solution is two full pages long, and it is extremely analytical and non-geometrical. He even commits the "heresy of Heron and Brahmagupta" and uses calculations that involve square roots of fourth powers to find areas. Orthodox geometers objected to such calculations because fourth powers had no "real" geometric interpretation. At this point in the paper, though, Euler is being true to his 18th century context, and uses fourth powers in geometry without agonizing over interpretation.

Having given a clunky proof to a somewhat awkwardly worded theorem, and used it to give a solution to a problem that holds little interest today, Euler spots a gem, and writes "the following most elegant consequence can now be stated:"

Theorem. *In an arbitrary triangle ABC, draw from each angle A, B and C to a point on its opposite side straight lines Aa, Bb and Cc so that the three segments intersect at some point O. Then the segments always have the property that*

$$\frac{Oa}{Aa} + \frac{Ob}{Bb} + \frac{Oc}{Cc} = 1.$$

Proof is a consequence of the previous theorem as follows.

As above, take $AO = \alpha \cdot Oa$, $BO = \beta \cdot Ob$, $CO = \gamma \cdot Oc$. The previous theorem gives that

$$\alpha\beta\gamma = \alpha + \beta + \gamma + 2.$$

Add $\alpha\beta + \alpha\gamma + \beta\gamma + \alpha + \beta + \gamma + 1$ to both sides. Then the left-hand side factors to give

$$(\alpha + 1)(\beta + 1)(\gamma + 1),$$

and the right-hand side can be written as

$$\alpha\beta + \alpha\gamma + \beta\gamma + 2(\alpha + \beta + \gamma) + 3.$$

This last "obviously" [Euler's word] resolves to give

$$(\alpha + 1)(\beta + 1) + (\alpha + 1)(\gamma + 1) + (\beta + 1)(\gamma + 1).$$

Now, dividing both sides by the product

$$(\alpha + 1)(\beta + 1)(\gamma + 1)$$

gives us

$$1 = \frac{1}{\alpha + 1} + \frac{1}{\beta + 1} + \frac{1}{\gamma + 1}.$$

Q.E.D.

Euler tells us that he's done with the proof even though this last formula is only equivalent to what we were trying to prove, after substituting geometric segments for the Greek letters.

This theorem is the starting point for a recent article by Grünbaum and Klamkin. [GK] They show, among other things, that the analogous sum of ratios also holds for tetrahedral, as well as for higher dimensional, simplices. They correctly credit Euler for this result in two dimensions, but they neglect to notice that Euler also proved the two-dimensional version of their Theorem 1 (ii). Euler doesn't put this particular result in the form of a theorem, but instead writes:

"Here is how the following memorable property can be derived:

$$\frac{\alpha}{\alpha + 1} + \frac{\beta}{\beta + 1} + \frac{\gamma}{\gamma + 1} = 2.$$

If this is added to the previous equation, it gives the following identity:

$$1 + 1 + 1 = 3."$$

Perhaps we should name this last equation the "Euler identity"? [1]

Note that Euler's three results so far, $\frac{\alpha}{\alpha+1} + \frac{\beta}{\beta+1} + \frac{\gamma}{\gamma+1} = 2$, $1 = \frac{\alpha}{\alpha+1} + \frac{\beta}{\beta+1} + \frac{\gamma}{\gamma+1}$ and $\alpha\beta\gamma = \alpha + \beta + \gamma + 2$ are algebraically equivalent. If we have a proof of any one of them, then the other two follow with just a little bit of algebra.

This seems as if it would have been a fine place to stop this paper. Perhaps Euler did stop here, because he continues with a proof of this last result that is so much nicer than the one he gave above. If he'd known it when he was writing the earlier part of the paper, he probably would have used this proof instead. Or perhaps he came back to the paper later and added this part. Regardless, he next gives us what he calls a "Most simple proof, based on ordinary elements." By this he means that he is going to use mostly geometry instead of mostly algebra to prove

$$\frac{\alpha}{\alpha+1} + \frac{\beta}{\beta+1} + \frac{\gamma}{\gamma+1} = 1.$$

Euler uses a new figure (Figure 2), and some new notation, and some new lines in his figure. Through O, he draws segments parallel to each of the three sides of $\triangle ABC$. The segment $f\zeta$ is parallel to BC, $g\eta$ is parallel to AC and $h\theta$ is parallel to AB. With this notation, Euler writes the property he plans to prove in the form

$$\frac{Oa}{Aa} + \frac{Ob}{Bb} + \frac{Oc}{Cc} = 1.$$

He begins his proof writing that since $AB = A\eta + \eta f + fB$, we have

$$\frac{Bf}{AB} + \frac{A\eta}{AB} + \frac{f\eta}{AB} = 1. \tag{5}$$

Now, since $\triangle ABa$ is similar to $\triangle AfO$, we get $Bf : BA = Oa : Aa$, or, as fractions,

$$\frac{Oa}{Aa} = \frac{Bf}{BA}. \tag{6}$$

Save this to substitute into (5).

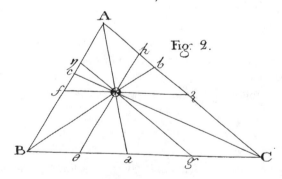

Mémoires de l'Académie Imp. des Sc. Tome V Tab. I.

Fig. 2.

Likewise from $\triangle BAb \sim \triangle B\eta O$ we get another ratio that makes

$$\frac{A\eta}{AB} = \frac{Ob}{Bb}. \tag{7}$$

The third of the similar triangles gives, in the same way $\frac{f\eta}{AB} = \frac{fO}{BC}$.

By parallel lines, $fO = B\theta$ and also $\triangle BCc \sim \triangle\theta CO$ so we get $\frac{B\theta}{BC} = \frac{Oc}{Cc}$. That makes

$$\frac{f\eta}{BA} = \frac{Oc}{Cc}. \tag{8}$$

Now we substitute formulas (6), (7) and (8) into the identity (5) $\frac{Bf}{AB} + \frac{A\eta}{AB} + \frac{f\eta}{AB} = 1$ and it gives our theorem

$$\frac{Oa}{Aa} + \frac{Ob}{Bb} + \frac{Oc}{Cc} = 1.$$

Q.E.D.

Euler adds that this property even holds if the point O is taken to be outside the triangle, as shown in Figure 3. Homer White, in footnotes to his translation of E749 (available on The Euler Archive) describes Euler's explanation of this property as "quite unclear." White also observes that the version of Figure 3 given in the *Opera Omnia* contains an error, reversing the labels on the points C and c. As we can see in our version of Figure 3, the points were correctly labeled in the original.

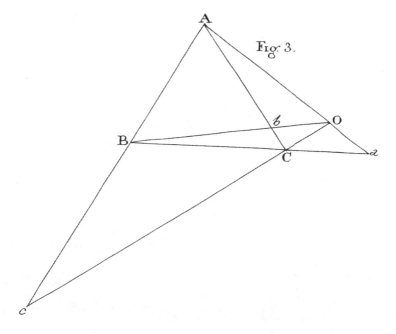

Fig. 3.

In typical Eulerian fashion, Euler proves that similar relations hold on coincident segments drawn across spherical triangles, as shown in Figure 4. He also shows how to solve his problem for spherical triangles. These calculations closely resemble the calculations we saw early in the paper, reinforcing the hypothesis that the "most simple proof based on ordinary elements" may have been added later.

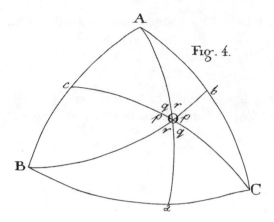

Fig. 4.

Whether or not that part was interpolated, the last part of this paper was clearly added later than the rest of the paper. Euler labels it "SUPPLEMENT Containing the simplest analysis for the proof of the theorem and for the solution of the problem proposed before."

Euler means to prove his theorem in the form $\frac{Oa}{Aa} + \frac{Ob}{Bb} + \frac{Oc}{Cc} = 1$.

For his proof, Euler uses Figure 5, a simpler version of Figure 2. Rather than add line segments parallel to all three sides, he adds only two shorter line segments, both from point O to points on side BC. Segment $O\beta$ is parallel to side AB and segment $O\gamma$ is parallel to AC. Obviously,

$$B\beta + \beta\gamma + \gamma C = BC,$$

So

$$\frac{B\beta}{BC} + \frac{\beta\gamma}{BC} + \frac{\gamma C}{BC} = 1. \tag{9}$$

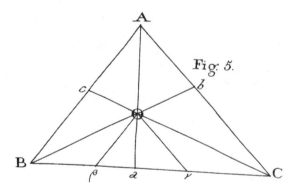

Fig. 5.

Now we use three pairs of similar triangles. $\triangle BCb \sim \triangle B\gamma O$ so $\frac{C\gamma}{BC} = \frac{Ob}{Bb}$. Likewise $\triangle CBc \sim \triangle C\beta O$ so $\frac{B\beta}{BC} = \frac{Oc}{Cc}$. Finally, $\triangle \beta O\gamma \sim \triangle BAC$, so $\frac{\beta\gamma}{BC} = \frac{Oa}{Aa}$. Substitute these three fractions into (9), and immediately we get

$$\frac{Oa}{Aa} + \frac{Ob}{Bb} + \frac{Oc}{Cc} = 1.$$

Q.E.D.

Euler is clearly proud of this proof and writes that "this is, without a doubt, the shortest proof of this theorem, but it was dug up in a most roundabout way."

He wraps up his paper with a not-quite-as-brief solution to his problem of finding the triangle given the segments. We note that, for this part, the illustration in the *Opera Omnia* again contains an error that was not present in the original.

So we reach the end of Euler's last published paper in Euclidean geometry. Because Euler didn't have the extra time to revise and polish this paper (he died just three years after writing it, and his SUPPLEMENT may have been added substantially closer to his death) he didn't "erase his tracks." Thus it gives us a glimpse of how Euler discovered things as he wrote a paper and how he came back later to improve his solutions.

It also shows how Euler, though nicknamed by one of his contemporaries *Analysis incarnate*, still had a flair for ordinary geometry and, though blind himself, still had an eye for a beautiful proof.

References

[CG]   Coxeter, H. S. M., and S. L. Greitzer, *Geometry Revisited*, Anneli Lax New Mathematical Library, MAA, Washington DC, 1967. (Originally published by Random House, New York, 1967.)

[E749] Euler, Leonhard, Geometrica et sphaerica quaedam, *Mémoires de l'Académie des sciences de St-Pétersbourg* 5 (1812) 1815, pp. 96–114 reprinted in *Opera Omnia* Series I vol. 26 pp. 344–358. Available in its Latin original and in translation by Homer White at EulerArchive.org.

[GK]   Grünbaum, Branko, and Murray S. Klamkin, Euler's Ratio-Sum Theorem and Generalizations, *Mathematics Magazine* 79:2 (April 2006) 122–130.

5

Beyond Isosceles Triangles

(April 2004)

We know lots about triangles for which $\angle A = \angle B$. Such triangles are isosceles, and we have known at least since Euclid that $\angle A = \angle B$ exactly when $a = b$. In 1765, Euler studied a generalization of this situation. What happens if $\angle B$ is some multiple of $\angle A$?

In 1765, Euler was about to return to St. Petersburg after 25 years in Berlin. His last years in Berlin had been rather unhappy, and he would be glad to be back in Russia, where the new Empress, Catherine II, also known as Catherine the Great, showered him with attention and honors. While he was in Berlin, he continued to edit the *Novi commentarii academiae scientiarum imperialis petropolitanae*, the main journal of the St. Petersburg Academy. He also published much of his most important work there. In 1766, Euler left Berlin and was back in St. Petersburg by the time the volume was actually published in 1767. In the 1765 volume, Euler wrote no less than ten papers on an amazingly diverse range of topics. One was about the nature of discontinuous functions, two about the three-body problem in celestial mechanics, one about the so-called Pell equation in number theory, one about friction in gears, and one was the paper in which he discovered the well-known Euler Line. Nestled among these others, numbered E324 in Eneström's index of Euler's work, was a 35-page paper, "*Proprietates triangulorum quorum anguli certam inter se rationem tenent*" (Properties of triangles for which certain angles have a ratio between themselves.)

Euler let ABC be a triangle. As usual, he denotes the side opposite angle A by a, opposite B by b and opposite C by c. To set up his problem, Euler supposes that the ratio between the angles is given by $\angle A : \angle B :: m : n$ and that $n = 1$. He considers various values of m. First, he reminds us that if $m = 1$, then we are talking about isosceles triangles, so $a = b$. To make this fit certain patterns later in the paper, he chooses to write this as $b - a = 0$. He continues to the case $m = 2$. By the end of the paper, he will have studied as far as $m = 13$.

In case $m = 2$, Euler poses the problem "to investigate the relations that arise." He gives us Figure 2 below. In Euler's day, all the figures in the book were printed together

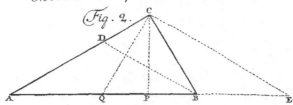

on pages bound at the back of the book, and numbered consecutively. Figure 1 was part of another article on discontinuous functions.

In Figure 2, Euler takes $\angle B = 2\angle A$ and draws BD bisecting $\angle B$. This makes $\angle A = \angle ABD = \angle DBC$ so triangle ADB is isosceles and triangles ABC and BDC are similar. He writes the similarities as

$$AC : BC = AB : BD = BC : CD$$

so $b : a = c : \frac{ac}{b} = a : \frac{aa}{b}$ and

$$BD = \frac{ac}{b}, \quad CD = \frac{aa}{b} \quad \text{and} \quad AD = b - \frac{aa}{b}.$$

Note that to make typesetting easier, in Euler's day they usually wrote aa where we would write a^2.

Euler has expressions for AD and for BD, and he knows they are equal since ADB is isosceles, and it follows immediately that $ac = bb - aa$. This is Euler's main result for triangles where one angle is the double of another.

Readers will note that Euler has only used about half of his diagram so far. He hasn't used the other points P, Q and E. He draws CP, an altitude of triangle ABC, and then locates Q and E so that $QP = PB$ and $AP = PE$. Then he uses this diagram to show that if the sides of a triangle satisfy the relation $ac = bb - aa$, then $\angle B = 2\angle A$. We will leave it to the reader to work that out.

Euler goes on to consider the case $m = 3$, where $\angle B = 3\angle A$. He uses Figure 3 below, in which he adds a new point c such that $\angle CBc = \angle A$. It doesn't bother Euler that c denotes both a point on side AC and the length of side AB because it is always clear what he means from the context. Since he already knows that $\angle B = 3\angle A$, he doesn't have to trisect $\angle B$ to do this, though the idea of finding a way to trisect angles probably occurred to him.

This leaves $\angle ABc = 2\angle A$, therefore his earlier results apply to triangle ABc. He writes the lengths of the sides of this triangle using Greek letters α, β and γ, so that

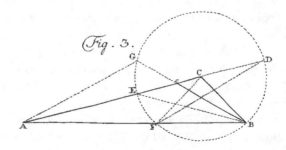

$\beta\beta - \alpha\alpha - \alpha\gamma = 0$. Also, since $\angle CBc = \angle A$, triangles ACB and BCc are similar. We then have the ratios $b : a = c : \frac{ac}{b} = a : \frac{aa}{b}$ and so

$$Bc = \frac{ac}{b}, \quad Cc = \frac{aa}{b} \quad \text{and} \quad Ac = \frac{bb - aa}{b}.$$

Comparing the sides that triangles ABc and ABC have in common, we know that $\gamma = c$, $\beta = \frac{bb-aa}{b}$ and $\alpha = \frac{ac}{b}$. Putting all this together and clearing the denominator, we get $(bb - aa)^2 - acc(a + b) = 0$. Then factoring out $a + b$ leaves

$$(bb - aa)(b - a) - acc = 0.$$

This is Euler's main result in the case $m = 3$.

As before, the remainder of Figure 3 plays a role in showing that triangles that satisfy the condition $(bb - aa)(b - a) - acc = 0$ must also satisfy $\angle B = 3\angle A$.

Euler continues this way, up to $m = 5$, each time drawing a line Bc to make $\angle CBc = \angle A$, and thus cutting off one triangle BCc similar to triangle ABC, and another triangle ABc for which the immediately previous results apply, that is for which $\angle CBc = (m - 1)\angle A$. He gets progressively more complicated relations among the sides, though he stops proving that triangles satisfying those relations have angles in the given ratios. With $m = 5$, Euler notices a pattern, proves that the pattern holds, and then uses the pattern to give relations up to $m = 13$.

We'll stop here, though, before we get buried in an avalanche of details. Though this certainly wasn't one of Euler's major results, it is a pleasant one, and it is a fine example of Euler's style.

References

[E324] Euler, Leonhard, Proprietates triangulorum, quorum anguli certam inter se rationem tenent, *Novi commentarii academiae scientiarum imperialis Petropolitanae* Vol. XI, 1765 (1767), pp. 67–102, reprinted in *Opera Omnia* Series I vol. 26 pp. 109–138.

6

The Euler–Pythagoras Theorem

(January 2005)

Euler didn't do a lot of geometry. Most of what he did falls into one of two categories. One category includes papers that were part of now-forgotten research agendas of the 1700s. Euler would usually do several papers on such a topic. His work on reciprocal trajectories and on the quadrature of lunes both fall into this category. The other category includes papers that are solitary gems in the field, topics Euler visited once, created a masterpiece, and then moved on. His work on the Euler Line and on the so-called Euler formula ($V - E + F = 2$, actually two papers) are examples here.

This month we look at a rare example from a third category of Euler's geometry, a topic he visited once and the single paper that resulted from that visit is mostly forgotten. It is a beautiful result that I learned from Bill Dunham at a meeting of the Ohio Section of the MAA. Though Euler did not bother to mention it, the Pythagorean theorem is an easy corollary of the main result, hence the title of this column.

In 1748 Euler had been in the employ of Frederick II in Berlin for seven years. He published one of his most influential works, the *Introductio in analysin infinitorum*, often heralded as the world's first precalculus book, though that description is a gross oversimplification. He finished writing his first calculus book, the *Institutiones calculi differentialis*, though it would not be published until 1755. He "only" published nine books and papers in 1748, but he wrote more than 25. Most of his papers were about astronomy, especially the motions of the moon and planets, and about optics and the tools necessary to observe those motions. Euler's paper, *Variae demonstrationes geometricae*, or "Several proofs in geometry," is a bit of an oddball in the midst of all this astronomy. It is number 135 on Eneström's list of Euler's works.

As its title suggests, *Variae demonstrationes geometricae* presents several results, not all that closely related, and not all that new. First, he solves a fairly simple and not very interesting problem of Fermat. Then he does two theorems in triangle geometry. The first of this pair is the theorem relating the area of a triangle to its perimeter and the radius of its inscribed circle. The second is the theorem often known as Heron's theorem, giving the area of a triangle in terms of the lengths of its three sides. He also proves what we

33

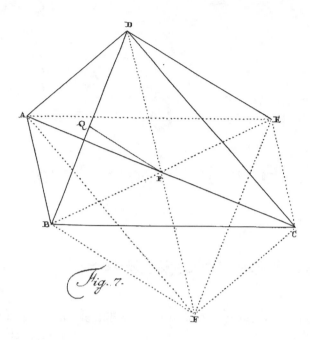

Fig. 7.

sometimes call Brahmagupta's theorem, giving the area of a quadrilateral inscribed in a circle in terms of the four sides of the quadrilateral. Euler attributes this theorem to one of his contemporaries, Philipp Naudé.

Finally, just three pages from the end of this 18-page article, he gets to the theorems that interest us today. Euler refers us to Figure 7 when he states the first of these theorems:

Theorem. *Given any convex quadrilateral (trapezio) $ABCD$ with diagonals AC, BD, if a parallelogram is completed about the two sides AB, BC to give the parallelogram $ABCE$, and if the two points D and E are joined to form the segment DE, then the sum of the squares of the four sides of the quadrilateral $AB^2 + BC^2 + CD^2 + DA^2$ will be greater than the sum of the squares of the diagonals $AC^2 + BD^2$ by the square of the segment DE, that is*

$$AB^2 + BC^2 + CD^2 + DA^2 = AC^2 + BD^2 + DE^2.$$

Proof. First, complete the three points A, B, C to form the parallelogram $ABCE$, as suggested in the wording of the theorem, and draw the diagonal BE. Further, construct F so that CF is parallel to AD and BF is parallel to ED. Since $BC = AE$, we get that the triangles CBF and AED are congruent.

Now, draw the lines AF, DF and EF and look at the two parallelograms $ADCF$ and $BDEF$ with diagonals AC, DF and BE, DF respectively. Euler cites a "property of parallelograms" to tell us that, in $ADCF$ we have

$$2AD^2 + 2CD^2 = AC^2 + DF^2$$

and in BDEF we have

$$2BD + 2DE^2 = BE^2 + DF^2.$$

This is just an application of the Law of Cosines. We know that angles ADC and DCF are supplementary, so their cosines are negatives of one another. Then the Law of Cosines tells us that $AC^2 = AD^2 + CD^2 - 2 \cdot AD \cdot CD \cdot \cos(ADC)$. Meanwhile, $DF^2 = AD^2 + CD^2 + 2 \cdot AD \cdot DC \cdot \cos(ADC)$. Add these together to get Euler's property of parallelograms.

Solve each of these equations for DF^2, set the two parts equal to each other and add AC^2 to get

$$2AD^2 + 2CD^2 = 2BD^2 + 2DE^2 + AC^2 - BE^2.$$

We have one parallelogram left, $ABCE$, where we know that

$$2AB^2 + 2BC^2 = AC^2 + BE^2.$$

Add this to the last equation to get

$$2AD^2 + 2CD^2 + 2AB^2 + 2BC^2 = 2BD^2 + 2DE^2 + 2AC^2.$$

Divide by two and rearrange a little bit to get

$$AB^2 + BC^2 + CD^2 + DA^2 = AC^2 + BD^2 + DE^2,$$

as promised.   Q. E. D.

Euler gives us four corollaries. The first three are fairly routine. First, if the quadrilateral is a parallelogram, then the interval DE vanishes so the sum of the squares of the sides exactly equals the sum of the squares of the diagonals. This isn't really fair, since this is exactly the "property of parallelograms" that Euler used to prove the theorem itself. Euler's second corollary is that the sum of the squares of the sides of a quadrilateral is always greater than or equal to the sum of the squares of the diagonals, with equality exactly when the quadrilateral is a parallelogram.

In Euler's third corollary he bisects diagonal AC at P and BD at Q, and draws segment QP. Then he shows that this last segment is half the length of DE, and so its square is one fourth of DE^2. This is satisfying because it gives us a way to avoid using that awkward auxiliary point E and to replace it with two points P and Q that are more naturally associated with the original quadrilateral.

Substituting $4PQ^2$ for DE^2 in the original theorem leads to Euler's fourth, and most interesting corollary, illustrated in Figure 8:

Corollary 4. *In any quadrilateral $ABCD$, if its diagonals AC and BD are bisected by points P and Q, which are joined by segment PQ, then the sum of the squares of the four sides, $AB^2 + BC^2 + CD^2 + DA^2$ is equal to the sum of the squares of the two diagonals, $AC^2 + BD^2$ plus four times the square of the line PQ. That is to say,*

$$AB^2 + BC^2 + CD^2 + DA^2 = AC^2 + BD^2 + 4PQ^2.$$

Euler stopped here, but we don't have to. In the special case that $ABCD$ is a rectangle, besides knowing that all four angles are right angles and that opposite sides are equal, we also know that $P = Q$ and $AC = BD$. Making appropriate substitutions, this tells us that in the right triangle ABC, we have

$$AB^2 + BC^2 + AB^2 + BC^2 = AC^2 + AC^2 + 4 \cdot 0,$$

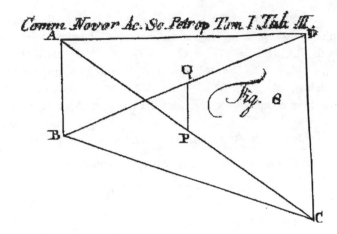

which gives immediately the well-known:

Theorem (Euler–Pythagoras). *If ABC is a triangle with right angle at B, then*

$$AB^2 + BC^2 = AC^2.$$

In the same volume of the *Novi Commentarii* as Euler published E135, his enthusiastic but less talented friend George Wolfgang Krafft (1701–1754) published an article, *Demonstrationes duorum theorematum geometricorum*, "Proofs of two geometric theorems." One of Krafft's theorems is Euler's Corollary 4, proved in an almost entirely algebraic manner, based on the Law of Cosines. Today, most of us probably find Euler's geometric method more appealing. But in the 1740s the fashion in mathematics was becoming more and more algebraic, at the expense of geometry. There is every chance that people in Euler's and Krafft's time found Krafft's the more attractive proof.

We leave the reader with two thoughts. First, how much of Euler's work is still true if the quadrilateral $ABCD$ is not convex, or is even self-intersecting?

The second thought is based on some remarks by Eisso Atzema. What if the four points A, B, C and D are not in a plane, so that they form the vertices of a tetrahedron. Then $ABCD$ describes a circuit on that tetrahedron. The two edges that the circuit does not use, AC and BD are the segments that are the diagonals in the planar case. How much of Euler's work survives this excursion into three dimensions?

References

[E135] Euler, Leonhard, Variae demonstrationes geometricae, *Novi commentarii academiae scientiarum Petropolitanae*, 1 (1747/48) 1750, pp. 49–66, reprinted in *Opera Omnia* Series I vol. 26 pp. 15–32. Available online at EulerArchive.org.

[K]   Krafft, George Wolfgang, Demonstrationes duorum Theorematum Geometricorum, *Novi commentarii academiae scientiarum Petropolitanae*, 1 (1747/48) 1750, pp. 131–136.

Illustrations were obtained from www.EulerArchive.org.

7

Cramer's Paradox

(August 2004)

Gabriel Cramer (1704–1752) is best known for Cramer's rule, a technique for solving simultaneous linear equations that is not very useful in practice, but is of immense theoretical importance in linear algebra. We will not be talking much about Cramer's rule here, though. Instead, we will discuss a question he asked of Euler, and how Euler answered it. That question got the name "Cramer's Paradox."

Cramer lived and worked in Geneva. One could do a compare-and-contrast using Euler and Cramer, both Swiss, but one spoke French at home, and the other German. Both earned graduate degrees very early with theses on sound, but one stayed in Switzerland while the other left and never returned. Cramer was born three years earlier, but died 31 years

Gabriel Cramer

earlier, in 1752 at the age of only 47. Euler and Cramer began corresponding in 1743. At least 16 of their letters survive (one was discovered as recently as 2003), the last one just two months before Cramer's death, and Cramer's son Philip continued the correspondence for another year.

Before we dive into Cramer's Paradox, we'd better make sure we have our definitions straight. Cramer and Euler were interested in real algebraic curves, curves defined by equations of the form $f(x, y) = 0$, where f is a polynomial of degree n. The special case $n = 3$ comes up later, so it is useful to note that third degree algebraic curves, what we call cubic curves, have the form

$$\alpha x^3 + \beta x^2 y + \gamma x y^2 + \delta y^3 + \varepsilon x^2 + \zeta x y + \eta y^2 + \theta x + \iota y + \kappa = 0.$$

There are ten coefficients to the cubic curve, the ten Greek letters given. Since for any nonzero constant c, $cf(x, y) = 0$ and $f(x, y) = 0$ define the same curve, we can pick any value we like (except zero) for one of the (nonzero) coefficients, and then the other nine coefficients are determined by the curve.

Several of the letters that Euler and Cramer exchanged dealt with questions about algebraic curves. At the time, mathematicians all believed, but were still unable to prove, what we now call the Fundamental Theorem of Algebra, that any polynomial of degree n has at most n real roots, and that, counting complex roots and multiplicities, it will have exactly n roots. Euler and d'Alembert both gave proofs that were accepted in the 1750s, but a proof that satisfies modern standards of rigor would have to wait for Gauss, in his doctoral thesis.

In one of those letters, dated September 30, 1744, Cramer asked the question that would soon bear his name, though some people trace its origins at least to Maclaurin, 15 or 20 years earlier. Cramer stated two "facts" about cubic curves:

1. One curve of order 3 (what we call *degree* he calls *order*, and hereafter we will try to use his term) can be determined by nine given points, and

2. Two curves of order 3 can intersect in nine points; thus one may find two curves of order 3 passing through these given nine points.

He asked how these can both be true, since the nine points of intersection seem to determine two different curves of order 3.

Euler worked on the question on and off for the next couple of years, and put his answer into an article that he wrote in 1747, *Sur une contradiction apparente dans la doctrine des lignes courbes* (On an apparent contradiction in the rule of curved lines). That article was published in 1748, and is number 147 on Eneström's index of Euler's works.

Euler begins his article with a "teaser." Though this is a real work of scholarship that addresses one of the important problems of the day, Euler shows a bit of a mischievous spirit in this and other passages throughout the paper:

> One believes generally that Geometry distinguishes itself from the other sciences because every advance is founded upon the most rigorous proofs, and that one will never find an occasion for controversy.

Of course, Euler is having fun here, since he plans to exploit a point where "Geometry" (what we call rigorous mathematics) was not quite rigorous at the time, and create what seems like a paradox or a controversy. He goes on to write

> I am going to describe two propositions of Geometry, both rigorously demonstrated, which would seem to lead to an open contradiction.

Euler restates Cramer's two "facts" and their generalizations. He reminds us that two points determine a line, a curve of order 1 (even though a line, in standard form $ax + by + c = 0$, has *three* coefficients, a, b and c). Five points determine a conic (order 2), and nine points determine a cubic. In general, Euler tells us that a curve of order n is determined by $(n^2 + 3n)/2$ points.

Euler calls this last claim Proposition 1. His proof is based on writing the general equation of a curve of order n as

$$Ay^n + (B + Cx)y^{n-1} + (D + Ex + Fx^2)y^{n-2} + (G + Hx + Ix^2 + Kx^3)y^{n-3} + \text{etc.} = 0.$$

This has

$$\frac{n^2 + 3n}{2} + 1$$

coefficients (the $(n + 1)$st triangular number, though not in its more familiar form, $(n + 1)(n + 2)/2$). Since one (nonzero) coefficient can be chosen arbitrarily (as noted above), the curve is determined by $(n^2 + 3n)/2$ linear equations. Since each given point provides us with an equation, we can determine the curve with $(n^2 + 3n)/2$ points.

If you see a flaw in this argument, then you might be able to see where this paper is going.

> Proposition 2: Two straight lines can cut each other in 1 point;
> Two conics can cut each other in 4 points.
> Two cubics can cut each other in 9 points.
> Two quartics can cut each other in 16 points.
> etc.
> A curve of order m can be cut by
> a line in m points,
> a conic in $2m$ points,
> a cubic in $3m$ points,
> and in general, an nth order curve in nm points.

He confesses that "The proof of this proposition is not so easy, and I will speak of it in more detail later in this discourse." Since the proof of this depends so much on the Fundamental Theorem of Algebra, and Euler can't prove that yet, he doesn't actually try to *prove* Proposition 2, but he does give several examples.

The problem is that Proposition 1 and Proposition 2 seem to contradict each other. Proposition 1 says, for example that nine points determine a cubic, and Proposition 2 says that two cubics can intersect each other at $3 \times 3 = 9$ points. So, pick two cubics that intersect in nine points. Use those nine points to determine a cubic. Which of the two cubics do the nine points determine?

As a concrete example, we could take the two cubics $x(x - 1)(x + 1) = 0$ and $y(y - 1)(y + 1) = 0$. The first gives three vertical lines, with x-intercepts -1, 0 and

1, and the second gives three horizontal lines. They intersect in the nine points with coordinates involving only -1, 0 and 1. Those nine points seem able to determine (at least) two different curves, $x(x-1)(x+1) = 0$ and $y(y-1)(y+1) = 0$.

The quandary gets even worse with quartics, where Proposition 1 tells us that 14 points determine a quartic, but Proposition 2 tells us that two quartics can intersect in 16 points, so 16 points seem to be able to determine *two* quartics. More constraints should give us *fewer* solutions, not more.

In the face of these apparent contradictions, Euler tells us that

it is absolutely necessary that either
a. One of the two general propositions is false, or
b. The consequences we draw are not justified.

Euler is working hard to build the dramatic tension here. Finally, without the modern tools of linear algebra, he gives a series of examples that work towards something like the rank of a system of linear equations. His first example is two lines:

$$3x - 2y = 5$$
$$4y = 6x - 10.$$

Here, two lines intersect, not at one point, but at infinitely many points. That might not be what Proposition 2 meant, since there are not really two lines here, just one line written two different ways. But he expands on the idea with a three-variable example,

$$2x - 3y + 5z = 8$$
$$3x - 5y + 7z = 9$$
$$x - y + 3z = 7.$$

Since the sum of equations 2 and 3 is the double of equation 1, these three equations have infinitely many solutions, not just one.

Euler further gives a four-variable example, and tells us that this can happen with any number of equations. He then uses the ideas to construct sets of points that correspond to more curves than Proposition 1 says they should. Let's look at what he does with a conic, or second order equation.

A conic is given by a second order equation like

$$\alpha x^2 + \beta xy + \gamma y^2 + \delta x + \varepsilon y + \zeta = 0.$$

Now, pick five points, denoted by Roman numerals I to V, such that

$$\text{I} = (0,0) \quad \text{II} = (a,0) \quad \text{III} = (0,b) \quad \text{IV} = (c,d) \quad \text{and} \quad \text{V} = (e,f).$$

Euler wants to find values of a, b, c, d and e so that these five points determine more than one conic curve. These five points give us five simultaneous equations:

$$\zeta = 0$$
$$\alpha a^2 + \beta 0^2 + \gamma 0^2 + \delta a + \varepsilon 0 + \zeta = 0$$
$$\alpha 0^2 + \beta 0^2 + \gamma b^2 + \delta 0 + \varepsilon b + \zeta = 0$$
$$\alpha c^2 + \beta cd + \gamma d^2 + \delta c + \varepsilon d + \zeta = 0$$
$$\alpha e^2 + \beta ef + \gamma f^2 + \delta e + \varepsilon f + \zeta = 0.$$

From the first three equations, we get

$$\zeta = 0$$
$$\delta = -\alpha a$$
$$\varepsilon = -\gamma b.$$

Substituting these into the last two equations, we get

$$\alpha c^2 + \beta cd + \gamma d^2 - \alpha ae - \gamma bd = 0$$
$$\alpha e^2 + \beta ef + \gamma f^2 - \alpha ae - \gamma bf = 0.$$

These two equations will determine the curve (two equations in three unknowns, α, β and γ, but we only need ratios), *unless* the two curves are equivalent. To find when they are equivalent, Euler solves both equations for β and sets them equal, to get

$$\beta = \frac{\alpha c(a-c) + \gamma d(b-d)}{cd} = \frac{\alpha e(a-e) + \gamma f(b-f)}{ef}.$$

A bit of algebra shows that, for these two expressions to be equal, it is sufficient that

$$a = \frac{cf - de}{f - d}, \quad \text{and} \quad b = \frac{de - cf}{e - c}.$$

So, if we pick our five points I, II, III, IV and V so that these last two relations hold, then we get five equations that have not just one, but infinitely many solutions, since the last two equations turn out to be equivalent.

We encourage the reader to experiment with this. For example, taking a, b, c, d, e and f to be -1, 1, 1, 2, 2 and 3 respectively, gives two distinct curves that have infinitely many points in common. Yet the points $(1, 1)$ and $(1, -1)$ are each on one of the curves, but not on the other.

This points the way to resolving the paradox. It is more complicated, but the same thing can be done with nine points and a cubic curve given by the equation

$$\alpha x^3 + \beta x^2 y + \gamma xy^2 + \delta y^3 + \varepsilon x^2 + \zeta xy + \eta y^2 + \theta x + \iota y + \kappa = 0.$$

We pick nine points, Euler would call them I to IX. We can assume a few of our coefficients are zero, but we are still left with a pretty nasty 9×9 system of equations. To add extra complication, there are a great many ways that the system might have infinitely many solutions. For example, equation VIII might be equivalent to equation IX, or equation VII might be equivalent to the sum of equations V and VI. Euler admits that "It is difficult to formulate a general case" that describes all the different ways the equation might have infinitely many solutions.

That would take linear algebra, a subject Euler helped make necessary, and which Cramer, with his Rule, helped discover.

So, there is no paradox, the scene is set for the invention of linear algebra, and mathematics survives another crisis.

References

[OC]   O'Connor, J. J., and E. F. Robertson, "Gabriel Cramer," The MacTutor History of Mathematics archive, `www-gap.dcs.st-and.ac.uk/~history/Mathematicians/Cramer.html`.

[E147] Euler, Leonhard, Sur une contradiction apparente dans la doctrine des lignes courbes, *Mémoires de l'Académie des Sciences de Berlin*, 4 1750, pp. 219–233, reprinted in *Opera Omnia* Series I vol. 26 pp. 33–45. Available online at `EulerArchive.org`.

Part II

Number Theory

8

Fermat's Little Theorem

(November 2003)

If p is a prime number and if p does not divide a, then $a^{p-1} \equiv 1 \pmod{p}$. This fact is sometimes known as Fermat's Little Theorem. There is a generalization: if $\varphi(n)$ is the number of positive integers less than n and relatively prime to n, and if a and n are relatively prime, then $a^{\varphi(n)} \equiv 1 \pmod{n}$. This is a generalization because n is prime exactly when $\varphi(n) = n - 1$. The generalization is sometimes known as the Euler–Fermat Theorem.

Fermat was not the first to discover Fermat's Little Theorem, and Euler was not the first to prove it. Both Fermat and Leibniz proved it first, though Euler seems to have been the first one to *publish* a proof. In fact, over the course of forty years, Euler published at least three different proofs; more than that depending on how "different" you think proofs have to be to be called "different." Here, we are going to look at the first of those three proofs, and at how Euler came to discover it.

Our story begins, appropriately, with Fermat, though not with Fermat's Little Theorem. While he described the result in a letter to Marin Mersenne in 1640 and he said he had a proof, he never published the result or the proof. Instead, we begin with a conjecture Fermat mentioned in several letters in the 1640s and 1650s, to Frenicle de Bessy, Pascal, Carcavi and others, claiming that all numbers of the form $F_n = 2^{2^n} + 1$ are prime. These numbers are now known as Fermat numbers. All of these people tried their hands at a proof. Frenicle de Bessy claimed a proof, but it does not survive.

Let us jump forward about sixty years to 1729, when a 22-year old Leonhard Euler works at the Imperial Academy of Sciences in St. Petersburg. Christian Goldbach (1690–1764) is Secretary of the Academy, and one of Euler's superiors. Euler and Goldbach exchange letters for more than 35 years. At the end of his very first letter of this long series, Goldbach remarks

P. S. Notane Tibi est Fermatiaii observatio omnes numeros hujus formulae $2^{2^x} + 1$, nempe 3, 5, 17, etc. esse primos, quam tamen ipse fatebatur se demonstrare non posse et post eum nemo, quod sciam, demonstravit.

P. S. Note the observation of Fermat that all numbers of this form $2^{2^x} + 1$, that is, 3, 5, 17, etc., are primes, which he himself admits that he was not able to prove, and, as far as I know, nobody else has proved it either.

Three years later, in a five-page paper that now bears the index number E26, Euler shows that the $F_5 = 4{,}294{,}967{,}297 = 641 \cdot 6{,}700{,}417$. That is, Fermat was wrong. At the end of the paper, Euler adds six "theorems" of his own, things he believes to be true but cannot prove. Yet. The first of these six "theorems" is exactly Fermat's Little Theorem, though the "(mod n)" notation has not yet been introduced. Three of the other "theorems" are special cases of the Euler-Fermat Theorem. At the time, Euler can't prove any of them.

By 1736, Euler makes great progress in discovering ways to prove theorems in number theory. In another five-page paper, [E54] he states and proves Fermat's Little Theorem. In describing his proof, we will use his notation, though some of it could be simplified a bit by using modern notation, and, where possible, we will use his words, though translated from Latin into English. He states the theorem thus:

Proposition. *With p signifying a prime number, the formula $a^p - 1$ will always be able to be divided by p, unless a itself is divisible by p.*

Euler plans to prove this by mathematical induction on a, but he does not expect his readers necessarily to be familiar with the technique, so he explains his steps carefully and breaks the steps into a number of lemmas. He feels it safe to omit the cases $p = 2$ and also $a = 1$, and begins with $a = 2$. He claims

Proposition. *With p signifying an odd prime number, then any formula $2^{p-1} - 1$ will always be able to be divided by p.*

For his proof, he wrote 2 as $1 + 1$ to get

$$(1 + 1)^{p-1} = 1 + \frac{p-1}{1} + \frac{(p-1)(p-2)}{1 \cdot 2} + \frac{(p-1)(p-2)(p-3)}{1 \cdot 2 \cdot 3}$$
$$+ \frac{(p-1)(p-2)(p-3)(p-4)}{1 \cdot 2 \cdot 3 \cdot 4} + \text{etc.}$$

where, Euler reminds us, the number of terms here is equal to p, which is an odd number. Note that Euler does not have a modern notation for the binomial coefficients, nor does he have the modern $\sum$- notation.

Next, Euler subtracts 1 from both sides, to get $(1 + 1)^{p-1} - 1$ on the left and an even number of terms on the right. Those terms on the right form pairs of consecutive binomial coefficients. Euler applies the well-known identity on binomial coefficients that we now write as

$$\binom{p-1}{k-1} + \binom{p-1}{k} = \binom{p}{k}.$$

This gives him

$$2^{p-1} - 1 = \frac{p(p-1)}{1 \cdot 2} + \frac{p(p-1)(p-2)(p-3)}{1 \cdot 2 \cdot 3 \cdot 4}$$
$$+ \frac{p(p-1)(p-2)(p-3)(p-4)(p-5)}{1 \cdot 2 \cdot 3 \cdot 4 \cdot 5 \cdot 6} + \text{etc.}$$

On the right, there are $\frac{p-1}{2}$ terms, the last of which is p. Each term on the right is divisible by p, so the left-hand side must also be divisible by p.   Q. E. D.

Euler offers an alternate proof that is simpler. He begins with

$$2^p = (1+1)^p = 1 + \frac{p}{1} + \frac{p(p-1)}{1 \cdot 2} + \frac{p(p-1)(p-2)}{1 \cdot 2 \cdot 3} + \cdots + \frac{p}{1} + 1$$

Subtracting $2 = 1 + 1$ from both sides leaves

$$2^p - 2 = \frac{p}{1} + \frac{p(p-1)}{1 \cdot 2} + \frac{p(p-1)(p-2)}{1 \cdot 2 \cdot 3} + \cdots + \frac{p}{1}.$$

Obviously, p divides the right-hand side, hence p divides $2^p - 2$. Since p is odd, it must divide $2^{p-1} - 1$.   Q. E. D.

Euler also proves the case $a = 3$ separately, calling it a theorem. He has called his earlier results propositions.

Theorem.   *With p denoting any prime number except 3, any formula $3^{p-1} - 1$ will always be able to be divided by p.*

Euler patterns his proof after his second proof of the case $a = 2$, and writes

$$3^p = (1+2)^p = 1 + \frac{p}{1} \cdot 2 + \frac{p(p-1)}{1 \cdot 2} \cdot 4 + \frac{p(p-1)(p-2)}{1 \cdot 2 \cdot 3} \cdot 8 + \cdots + \frac{p}{1} \cdot 2^{p-1} + 2^p.$$

Now, Euler subtracts $1 + 2^p$ from both sides, and juggles the left-hand side a bit to get

$$3^p - 2^p - 1 = 3^p - 3 - 2^p + 2 = (3^p - 3) - (2^p - 2).$$

This leaves a factor of p in every term of the right-hand side, hence p must divide the right-hand side. The previous theorem showed that p divides $2^p - 2$, hence p must also divide $3^p - 3$. Since p is not 3, this leaves that p divides $3^{p-1} - 1$.   Q. E. D.

Euler is now ready to do his general induction step.

Theorem.   *With p denoting any prime number, if $a^p - a$ can be divided by p, then that same prime p will divide any formula $(a + 1)^p - a - 1$.*

Euler's calculations are identical to those in the case $a = 3$, with the 3's replaced with a's.

For his readers who are not familiar with mathematical induction, Euler concludes by explaining why the hard work is finished. His explanation is complete, but not all that clear. He showed that if $a^p - a$ is divisible by p, then so also is $(a + 1)^p - a - 1$. It follows that $(a + 2)^p - a - 2$ is also divisible by p, as is $(a + 3)^p - a - 3$, and, in general $(a + b)^p - a - b$. Since the theorem is true for $a = 2$, then any formula of the form $(b + 2)^p - b - 2$, whatever value is substituted for b. This shows that $a^p - a$ is always divisible by p.

Euler finishes with a brief remark that this implies that if p does not divide a, then p divides $a^{p-1} - 1$, as promised.

Fermat's Little Theorem has been proved in many ways. Probably only the Pythagorean Theorem has been proved in more ways. Over the course of his career, Euler himself will give at least two other proofs, and those may be topics for future columns.

As a final note, Euler refers, at various times, to at least ten different results as "a theorem of Fermat," but the result we now call "Fermat's Little Theorem" was not one of them. When he first discovered the theorem in 1732, he apparently did not know of Fermat's work on the subject.

References

[E26] Euler, Leonhard, Observationes de theoremate quodam Fermatiano aliisque ad numeros primos spectantibus, *Commentarii academiae scientiarum Petropolitanae* 6, (1732/3) 1738, pp. 103–107, reprinted in *Opera Omnia* Series I vol. 2 pp. 1–5. Available online at `EulerArchive.org`.

[E54] ——, Theorematum quorundam ad numeros primos spectantium demonstratio, *Commentarii academiae scientiarum Petropolitanae* 8, (1736) 1741, pp. 141–146, reprinted in *Opera Omnia* Series I vol. 2 pp. 33–35. Available online at `EulerArchive.org`.

9

Amicable Numbers

(November 2005)

Six is a special number. It is divisible by 1, 2 and 3, and, in what at first looks like a strange coincidence, $6 = 1 + 2 + 3$. The number 28 shares this remarkable property; its divisors, 1, 2, 4, 7 and 14, sum to the number 28. Numbers with this property, that they are the sum of their divisors (including 1, but not including the number itself) have been known since ancient times and are called *perfect numbers*. Euclid himself proved in Book IX, proposition 36 of the *Elements* [E]:

> If as many numbers as we please beginning from a unit be set out continuously in double proportion until the sum of all becomes prime, and if the sum multiplied into the last make some number, then the product will be perfect.

In a more modern treatment, Hardy and Wright [HW] state this same theorem as

Theorem 276.   *If $2^{n+1} - 1$ is prime, then $2^n(2^{n+1} - 1)$ is perfect.*

Each such perfect number is associated with a prime of the form $2^{n+1} - 1$, and such numbers are now called *Mersenne primes*. Several Mersenne primes are known, and for several decades, the largest known prime number was usually a Mersenne prime. This is no longer the case.

Euler proved that all even perfect numbers have the form in Theorem 276, and also discovered a few properties that a perfect number would have to have if it were odd.[1] Since no odd perfect numbers are known, it is difficult to explain to non-mathematicians why it might be interesting to prove things about them anyway. As far as I know, the two best-known properties of odd perfect numbers are:

1. There might not be any, and

2. if there are any, they must be very large.

But we are off the track of the story. Consider the pair of numbers, 220 and 284. The divisors of 220 are 1, 2, 4, 5, 10, 11, 20, 22, 44, 55 and 110, and those divisors sum to

---

[1] A year later, in November 2006, we returned to the properties of odd perfect numbers. See the next column in this collection.

284. Meanwhile, the divisors of 284 are 1, 2, 4, 71 and 142, and they sum to 220. Such pairs of numbers, the divisors of one summing to the other, are called *amicable pairs*.

For over a thousand years, only this pair, 220 and 284, was known. Iamblichus, in the fourth century BCE, wrote, "The first two friendly numbers are these: sigma pi delta and sigma kappa." In the Greek number system in use at the time, sigma had a value 200, pi and kappa were 80 and 20 respectively, and delta was 4, so he was describing 284 and 220.

In the 9th century, Arab mathematician Thabit ibn Qurra probably discovered the next amicable pair, 17296, 18416. In the 1600s, Pierre Fermat rediscovered this pair, and his mathematical rival René Descartes discovered another pair, 9,363,584 and 9,437,056.

So, when Euler came on the scene, only three pairs of amicable numbers were known. Then, in 1747, Euler published a short paper [E100] mentioning the technique that Descartes and Fermat had used, and listing 30 amicable pairs, including the three already known, and including one "pair" that was not actually amicable. Nevertheless, in one paper, Euler lengthened the list of known amicable pairs by a factor of almost ten.

Euler gives us almost no clue about how he found these numbers. He briefly describes the methods Descartes and Fermat had used, though. They had considered pairs of numbers of the form $2^n xy$ and $2^n z$, where x, y and z are all prime, and showed that, for the numbers to be an amicable pair, it was necessary that $z = xy + x + y$. Fermat and Descartes had just searched for prime numbers x, y and z to see which ones gave amicable pairs.

However, this cannot be how Euler found his new amicable pairs, since only the first three, the ones that were already known, have this form. Eleven of the others have the form $2^n xy$ and $2^n zw$, where x, y, z and w are all prime, but others involve as many as seven distinct prime factors, and ten of the pairs are pairs of odd numbers.

It is not like Euler to leave us in the dark like this, without showing us how he made his discoveries, and I can offer no very satisfying explanation. It is true that most articles published in the *Nova acta eruditorum* were rather brief, but this article was only three pages long. An author of Euler's stature would have been welcome to write six or seven pages, if he had wanted to. It is also true that few important mathematicians had worked on number theory since the days of Fermat, who died in 1665, 80 years before Euler wrote this article, and this was only Euler's sixth article that the Editors of the *Opera Omnia* classify as "number theory." Since Euler published over 90 such articles, E100 comes quite early in his number theory career. None of this seems to explain why Euler chose to be so obscure.

Later in 1747, though, Euler wrote another paper, *Theoremata circa divisores numerorum*, or "Theorems about divisors of numbers," [E134] in which he explained how he had discovered that the fifth Fermat number, $2^{2^5} + 1$, was not prime but was divisible by 641, and also gave his first proof of Fermat's Little Theorem. This was the subject of the very first column in this series, back in November of 2003. Perhaps that paper got Euler thinking about providing better explanations of his discoveries in number theory, or maybe it just kept him interested in number theory.

Whatever the reason, in 1750, Euler returned to the problem of amicable numbers, armed with a powerful new idea, the first of what we now call *number theoretical functions*. He invented a new function and a new notation, denoting the sum of the divisors of a number n, including n itself, by $\int n$. The integral sign is supposed to remind us that we are summing something. This function is now sometimes called the *sigma function* and denoted $\sigma(n)$. Here we will use Euler's notation.

Immediately after introducing his new notation, Euler gives the example that $\int 6 = 1+2+3+6 = 12$, and that, in general, perfect numbers are those for which $n = \int n - n$ and prime numbers are those for which $\int n = 1 + n$. He pays due attention to his fundamental case, $\int 1 = 1$, and notes that this shows that "the unit ought not be listed among the prime numbers."

He follows with an exposition almost indistinguishable from that in a modern number theory textbook, of the basic properties of his new function:

Lemma 1. *If m and n are relatively prime, then*

$$\int nm = \int m \cdot \int n.$$

Corollary. *If m, n and p are prime numbers, then*

$$\int mnp = \int m \cdot \int n \cdot \int p = (1 + m)(1 + n)(1 + p).$$

Lemma 2. *If n is a prime number, then*

$$\int n^k = 1 + n + n^2 + \cdots + n^k = \frac{n^{k+1} - 1}{n - 1}.$$

Lemma 3. *If a number N has a prime factorization $N = m^a \cdot n^\beta \cdot p^\gamma \cdot q^\delta \cdot$ etc.,* then

$$\int N = \int m^\alpha \cdot \int n^\beta \cdot \int p^\gamma \cdot \int q^\delta \cdot \text{etc.}$$

Euler does a few examples like finding that $\int 360 = 1170$ and using his new function to show that 2620 and 2924 form an amicable pair. With this last example, he is showing off a bit, since this pair is not among the three amicable pairs known in ancient times, though it was on his list in E100. Then he turns to characterizing amicable numbers. Here is how he does it.

If m and n are amicable pairs, then $\int m - m = n$ and $\int n - n = m$, and a little bit of algebra leads to the form Euler wants: $\int m = \int n = m + n$. Armed with this, he begins to study amicable pairs that share a common factor, a. He classifies these as follows:

$$\text{first form} \begin{cases} apq \\ ar \end{cases} \quad \text{second form} \begin{cases} apq \\ ars \end{cases}$$

$$\text{third form} \begin{cases} apqr \\ as \end{cases} \quad \text{fourth form} \begin{cases} apqr \\ ast \end{cases} \quad \text{fifth form} \begin{cases} apqr \\ astu \end{cases}$$

A modern reader might want to count the factors that the pairs do not have in common, and then classify these with a notation like $(2, 1), (2, 2), (3, 1), (3, 2), (3, 3)$, etc. We could then try to make a case that it resembles Cantor's diagonal proof that the rational numbers are countable, but such observations are anachronistic, and are more amusing than they are useful or valid.

Now he considers these one form at a time.

Problem I

First, Euler considers amicable pairs of the form apq and ar, where there is a common factor a and the numbers p, q and r are prime numbers and not factors of a. All of the amicable pairs known before Euler's time were of this form and had a being a power of 2.

The condition he found earlier implies that $\int r = \int p \cdot \int q$, and, since p, q and r are prime, this means that $r + 1 = (p + 1)(q + 1)$.

Substituting x for $p + 1$ and y for $q + 1$, this makes $r = xy - 1$, where the numbers $x - 1$, $y - 1$ and $xy - 1$ must all be prime, and the numbers $a(x - 1)(y - 1)$ and $a(xy - 1)$ form the amicable pair he is seeking. Moreover, the condition $\int m = \int n = m + n$ becomes

$$a(2xy - x - y) = xy \int a \quad \text{or} \quad y = \frac{ax}{(2a - \int a)\, x - a}.$$

He simplifies this with the substitution

$$\frac{b}{c} = \frac{a}{2a - \int a},$$

$\frac{b}{c}$ taken to be in lowest terms. Substituting this into the expression for y, he gets the fairly simple form

$$(cx - b)(cy - b) = bb$$

and, because p, q and r are prime, he gives the additional conditions that $x - 1$, $y - 1$ and $xy - 1$ must all be prime.

This is enough new information to start searching for amicable pairs. He begins what he calls Rule 1, and supposes that a is a power of 2, say $a = 2^k$. His substitutions lead to $b = 2^n$ and $c = 1$, so that

$$(x - 2^n)(y - 2^n) = 2^{2n}.$$

Euler didn't leave out any steps of this calculation, but in Euler's day, paper was expensive, and we have a choice of cheap paper or computer algebra systems if we want to check his work.

Continuing, there aren't very many ways to factor 2^{2n}, and this product must have the form

$$(x - 2^n)(x + 2^n) = 2^{n+k} \cdot 2^{n-k}$$

for some value of k. From this it follows that

$$x = 2^{n+k} + 2^n$$
$$y = 2^{n-k} + 2^n$$

and the three prime numbers p, q and r that go into making the amicable pair are

$$p = x - 1 = 2^{n+k} + 2^n - 1$$
$$q = y - 1 = 2^{n-k} + 2^n - 1$$
$$r = xy - 1 = 2^{2n+1} + 2^{2n+k} + 2^{2n-k} - 1.$$

Euler makes one more, in this case rather unnecessary substitution, taking $m = n - k$, so that $n = m + k$, and rewrites these equations in terms of m and k instead of n and k. We'll skip that.

Now, Euler considers as separate cases various values of k.

First, if $k = 1$, we look for primes of the forms

$$p = 3 \cdot 2^m - 1$$
$$q = 6 \cdot 2^m - 1$$
$$r = 18 \cdot 2^{2m} - 1.$$

If $m = 1$, then these give prime numbers 5, 11 and 71, and so the numbers

$$220 = 2^2 \cdot 5 \cdot 11$$
$$284 = 2^2 \cdot 71$$

form an amicable pair.

If $m = 2$, we get the numbers 11, 23 and 287. The first two are prime, but the third is 7×41, and so this case does not yield an amicable pair.

If $m = 3$, we get the numbers primes 23, 47 and 1151, and hence the amicable pair

$$17,296 = 2^4 \cdot 23 \cdot 47$$
$$18,416 = 2^4 \cdot 1151$$

After a few more unfruitful substitutions for $m = 4$ and $m = 5$, taking $m = 6$ gives another amicable pair, which we will leave to the reader to calculate.

Euler also considers cases $k = 2, 3, 4$ and 5, but they yield no additional amicable pairs. He assures us that these three are the only amicable pairs of this first form involving a common factor of 2^n and involving only prime numbers less than 100,000.

He further considers, without any positive results, common factors of the form $a = 2^n(2^{n+1} + 2^k - 1)$, for which the second factor is also a prime number. Euler calls this second prime factor f, and with a calculation almost exactly like the one above, concludes that for an amicable pair to be generated, there must be exponents m and n, for which $x = 2^n + 2^{n+1-m}$ and $y = (2^{n+1} + 2^{m+n} - 1)(2^n + 2^{n+1-m})$ for which $m < n + 1$ and all four of the following numbers must be prime:

$$f = 2^{n+1} + 2^{m+n} - 1$$
$$p = x - 1$$
$$q = y - 1$$
$$r = xy - 1$$

This is as far as Euler can go here with analysis, so it is time to examine cases. Taking $m = 1$ yields no amicable pairs.

However, if $m = 2$, it makes

$$f = 3 \cdot 2^{n+1} - 1, \quad x = 3 \cdot 2^{n-1}, \quad y = 3 \cdot 2^{n-1}(3 \cdot 2^{n+1} - 1), \quad \text{and} \quad a = 2^n \cdot f$$

whence

$$p = 3 \cdot 2^{n-1} - 1, \quad q = 3 \cdot 2^{n-1}(3 \cdot 2^{n+1} - 1) - 1 \quad \text{and} \quad r = 9 \cdot 2^{2n-2}(3 \cdot 2^{n+1} - 1) - 1.$$

One need only substitute various values of n, hoping to make all four of the numbers f, p, q and r prime. Euler does this in a table:

$n =$	1	2	3	4	5
$f =$	11	23	47	95*	191
$p =$	2	5	11	$\cdots$	47
$q =$	32*	137	563	$\cdots$	9167*
$r =$	98*	827	6767*	$\cdots$	$\cdots$

In this table, numbers that are not prime are marked with a *, and the ellipses mark numbers that were unnecessary to calculate because there is already a composite number in that column. The 98 in column 1 was unnecessary, but easy, so Euler did it anyway.

Only column 2 is free of composite numbers, and this means that it leads to a new amicable pair:

$$\begin{cases} 4 \cdot 23 \cdot 5 \cdot 137 \\ 4 \cdot 23 \cdot 827 \end{cases}$$

This is the first new amicable pair that Euler has shown us how to find, and it was the first new one on his list back in E100.

After this, the fun is over, even though the paper is less than half finished. Euler continues for another 50 pages, doing more forms, more cases, and turning up more and more amicable pairs. At the end of the paper, he summarizes his results, giving 61 amicable pairs, with a couple of typographical errors and a couple of mistakes, and doubling again the world's population of known amicable numbers.

Rather than slog through this, we'll leave it to the interested reader.

Euler wrote a third article with this same title, *De numeris amicabilibus* [E798]. He didn't finish it and it was not published during his lifetime, but was found among his papers and published in 1849, more than 60 years after his death. It is more pedagogical than the other two papers, and he does not give any new amicable pairs.

Amicable numbers are a curious topic. Euler's methods succeeded in reducing a nearly impossible search for special pairs of numbers to a more manageable search. He couldn't guarantee that numbers of a particular form would be amicable, but he made the search small enough that he was able to find quite a few of them. Now, using Euler's methods, but using computers to do the gigantic calculations, thousands of amicable pairs are known.

References

[E]   Euclid, *The Thirteen Books of Euclid's Elements*, 2ed, Sir Thomas Heath, tr., Cambridge University Press, 1926. Also available as a Dover reprint.

[HW]   Hardy, G. H., and E. M. Wright, *An Introduction to the Theory of Numbers*, 3ed, Oxford University Press, 1954.

[E100] Euler, Leonhard, De numeris amicabilibus, *Nova acta eruditorum*, 1747, pp. 267–269, reprinted in the *Opera Omnia*, Series I vol. 2, pp. 59–61. Also available online at www.EulerArchive.org.

[E134] ——, Theoremata circa divisores numerorum, *Novi commentarii academiae scientiarum Petropolitanae*, 1 (1747/48), 1750, pp. 20–48, reprinted in the *Opera Omnia*, Series I vol. 2, pp. 62–85. Also available online at www.EulerArchive.org.

[E152] ——, De numeris amicabilibus, *Opuscula varii argumenti* 2, 1750, pp. 23–107, reprinted in the *Opera Omnia*, Series I vol. 2, pp. 86–162. Also available online at www.EulerArchive.org.

[E798] ——, De numeris amicabilibus, *Commentationes arithmeticae* 2, 1849, pp. 627–636, reprinted in the *Opera Omnia*, Series I vol. 4, pp. 353–365. Also available online at www.EulerArchive.org.

10

Odd Perfect Numbers

(November 2006)

The subject we now call "number theory" was not a very popular one in the 18th century. Euler wrote almost a hundred papers on the subject, but the first book to be published on the subject seems to be Legendre's *Essai sur la théorie des nombres*, [L] published during "*an* VI," the sixth year of the French Revolutionary calendar, known to the rest of the world as 1798. Gauss's great *Disquisitiones arithmeticae* [G] followed just three years later.

Of course, like many questions people ask in the history of mathematics, if you don't like the answer to a question, you can interpret the question a little differently and get a different answer. Euclid's books VII to IX, [E] for example, are devoted to what we now call number theory. Here we find such essential theorems as the Euclidean Algorithm for finding the greatest common divisor [Book VII prop. 2], the theorem that there are infinitely many prime numbers [Book IX prop. 20] and, as the climax to Book IX, the theorem that if a number of the form $2^n - 1$ is prime, then the number $2^{n-1}(2^n - 1)$ is what he (and we) call a *perfect* number. [Prop. 36]

A case could also be made that the rare and obscure book *Exercitationum mathematicarum libri quinque*, by Frans van Schooten, (1615–1660) [S] published in 1656, was also a number theory book, since van Schooten wrote the book to explain how to find *amicable* numbers. That book did not have the impact van Schooten had hoped. People read his collected works of Viète and his Latin edition of Descartes' *Geometria*, but when they look at his *Exercitationum*, it is usually to read the appendix, *Tractatus de ratiociniis in aleae ludo*, by the young Christiaan Huygens, and a landmark in the history of probability. That, though, is a topic for another time and place.

It should be no surprise that a case could be made that Euler wrote, but never published, the first number theory textbook. It was a little-known manuscript, *Tractatus de numerorum doctrina capita sedecim quae supersunt*, "Tract on the doctrine of numbers, consisting of sixteen chapters." This is an unfinished first draft of part of what Euler apparently planned to be a textbook on number theory. He certainly did not intend it to be published in this form, and it was only published in 1849, 66 years after his death. We are not sure when Euler did this work. He probably wrote it after 1756, because it contains results that he published around that time. Most Euler experts agree that the manuscript was written before he died in 1783, but it is difficult to be more specific than that. André Weil [W] has some thoughts on the subject.

As the title makes clear, the *Tractatus de numerorum* consists of sixteen chapters, the titles of which are listed below:

Ch 1   On the composition of numbers

Ch 2   On the divisors of numbers

Ch 3   On the number of divisors and their sum

Ch 4   On relatively prime numbers and composites

Ch 5   On remainders born of division

Ch 6   On the remainders arising from the division of terms of an arithmetic progression

Ch 7   On the remainders arising from the division of terms of a geometric progression

Ch 8   On the powers of numbers which are left by the division by a prime number

Ch 9   On the divisors of numbers of the form $a^n \pm b^n$

Ch 10   On the remainders arising from the division of squares by prime numbers

Ch 11   On the remainders born of the division of cubes by prime numbers

Ch 12   On the remainders arising from the division of fourth powers (*biquadratorum*) by prime numbers

Ch 13   On the remainders arising from the division of fifth powers (*surdesolidorum*) by prime numbers

Ch 14   On the remainders arising from the division of squares by composite numbers

Ch 15   On the divisors of numbers of the form $xx + yy$

Ch 16   On the divisors of numbers of the form $xx + 2yy$

From the titles of the chapters, it appears that Euler's major interest in starting to write this book is to study the prime divisors of certain binary forms. It is likely that he had planned more chapters, perhaps one titled "On the divisors of numbers of the form $xx + nyy$." Euler wrote papers on this and related topics in the 1740s and 1750s. We have devoted earlier columns to two of these papers, one in December 2005 and another in January 2006.

Gauss, [G] in contrast, clearly had different plans when he wrote his *Disquisitiones arithmeticae*. His table of contents shows that he thought that the highlights were the Quadratic Reciprocity Theorem in Chapter 4 and the construction of the 17-gon in Chapter 7. (His preface also suggests that he didn't do all he had planned, as he mentions the contents of a Chapter 8 that doesn't exist.) Alas, I haven't had the chance to study Legendre's book, [L] so I can only guess what he intended.

Back to Euler's book. Euler starts with quite elementary material, giving "definitions" of numbers, arithmetic sequences, multiples, prime numbers, and other basic objects. He takes special care to note that a number na is the nth multiple of the number a, and also the ath multiple of the number n, and that $an = na$. In paragraph 32 (of 586 paragraphs) he tells us that prime numbers are those "numbers that are not multiples of any other number," and includes 1 among the primes when he first lists them, but later in the manuscript, he never seems to treat 1 as a prime.[1] If Euler had ever made a second draft of this manuscript, this is the kind of detail that he certainly would have straightened out.

Euler soon launches into a classification of numbers into the number of (not necessarily distinct) prime factors they have. Prime numbers are of the "first class." Squares and products of two primes are of the "second class," which includes numbers like 4, 6, 9, 10 and 14, but not numbers like 12. Since $12 = 2 \cdot 2 \cdot 3$, it is of the third class. He gives us the class (and prime factorization) of every number up to 100, and notes that 64 and 96 are the only numbers less than 100 of the sixth class.

We never do this today, but Euler, as usual, has a good reason and a good idea. He classifies the number of each class into *species*, according to the frequencies of the prime numbers in the prime factorization of a number. A number of the second class, for example, might be of the *first species*, having a form pp, like 4 or 9, or it might be of the *second species*, having form pq, like 6 or 15. Similarly, the species of the third class are p^3, p^2q and pqr. He surely recognizes, but does not mention, how the number of species of a class is related to the number of partitions of the class number.

We never see such a classification in modern approaches, though something similar occurs both when van Schooten [S] and when Euler do their analyses searching for amicable numbers. Here, though, it makes it easy for Euler to explain that the number of divisors of a number $n = p^\lambda q^\mu r^\nu s^\xi$ will be $(\lambda + 1)(\mu + 1)(\nu + 1)(\xi + 1)$. This function is called a *divisor function*[2] and is sometimes denoted either $d(n)$ or $\sigma_o(n)$. [Anon]

From there, it is an easy transition to the topic of Euler's chapter 3, "On the number of divisors and their sum." Euler turns to the sum of the divisors of a number n, a sum he denotes using an integral sign, $\int n$, but we now denote either $\tau(n)$ or $\sigma_1(n)$, the so-called "Euler tau function." With Euler's preparation, it is easy for him to show that for powers of prime numbers,

$$\int p^n = \frac{p^{n+1} - 1}{p - 1},$$

and if we know the prime factorization of a number to be $n = p^\lambda q^\mu r^\nu s^\xi$, then

$$\int n = \int p^\lambda \int q^\mu \int r^\nu \int s^\xi.$$

Euler proves a couple of lemmas about this function, for example that $\int n > n$, and if $n = 1$, then $\int 1 = 1$. Euler is just a little bit confusing here in his use of the ">" sign. We never see Euler using a $\geq$ sign, and here it looks like maybe he should have. He also finds $\int n$ for all n up to 60.

---

[1] I tell my students that the notion of "prime number" came from the ancient Greeks, who did not regard 1 as a number at all. This interpretation survives in modern English usage in our use of the phrase "a number of ..." If you ask me "How many sisters do you have?" and I reply, "I have a number of sisters," then you feel I have misled you with word games when you learn that I have only one sister. The phrase "a number of..." usually means "two or more," or even "three or more."

[2] *Wikipedia* [Anon] tells us that a *general divisor function* $\sigma_x(n)$ is the sum of the xth powers of the divisors of n, and can be denoted by $\sigma_x(n) = \sum_{d \mid n} d^x$.

Among his lemmas is that if $\frac{\int N}{N} = \frac{m}{n}$, with $\frac{m}{n}$ in lowest terms, if N is not equal to 1, and if $N \neq n$, then $m > \int n$. So, if $m = \int n$, then we must have $N = n$. It is not possible for $m < \int n$ since $\int N > N$.[3]

In paragraph 106 (still part of chapter 3) Euler defines a number N to be *perfect* if $\int N = 2N$. This is slightly different from the way Euclid [E] defines perfect numbers. For Euclid, N is perfect if it is exactly the sum of its divisors, (not twice that sum), but for Euclid, a number does not *divide* itself, because when you divide N into parts each of size N, you don't get any parts; you get the whole thing.

Euler neglects to point out that among the examples he'd done in earlier paragraphs, he had shown that $\int 6 = 12$ and $\int 28 = 56$, so both 6 and 28 are perfect numbers. Perhaps he would have added this in a later draft of the manuscript, or maybe he expected his readers to notice that. Euler begins to look for perfect numbers.

First, he supposes that N is even, so that $N = 2^n A$, for some power of 2 and some odd number A. Then

$$2^{n+1} A = 2N = \int N = \int 2^n A = (2^{n+1} - 1) \int A,$$

which makes

$$\frac{\int A}{A} = \frac{2^{n+1}}{2^{n+1} - 1}.$$

On the right, we have a fraction where the numerator is only one more than the denominator, hence it is in lowest terms. Now, if we take

$$\frac{2^{n+1}}{2^{n+1} - 1} = \frac{m}{n}$$

in lowest terms, as in the lemma, then either $m > \int n$ or $m = \int n$. The first case, Euler tells us, "yields no solutions." In the second case, the lemma tells us that $A = 2^{n+1} - 1$. Hence $\int A = A + 1$ and A is a Mersenne prime. Hence, any *even* perfect number has the form $2^n A$ where $A = 2^{n+1} - 1$ and A is prime. This is the converse to Euclid's Book IX prop. 36, and is the first proof of the converse.

Now we turn to the results that give this column its title, Euler's work on odd perfect numbers. Suppose that N is an odd perfect number and that it factors to be $N = ABCD$ etc. Euler implicitly assumes that these factors, A, B, C, D, etc. are powers of distinct primes, though he does not say so explicitly. Since N is odd, all of these factors must be odd, and since the prime factors are distinct, it must be that

$$2N = \int N = \int ABCD \text{ etc.} = \int A \int B \int C \int D \text{ etc.}$$

This number $2N$ is the double of an odd number, what Euler and Euclid called "oddly even" and we would call "congruent to 2 mod 4." Hence, it is divisible by 2 and not by 4, and so among its factors, $\int A$, $\int B$, $\int C$, etc., there must be one that is oddly even, and all the rest must be odd.

Suppose that one of them, say B, has $\int B$ odd, and that $B = p^n$. From what Euler did a long time ago,

$$\int p^n = \frac{p^{n+1} - 1}{p - 1} = p^n + p^{n-1} + \cdots + p^2 + p + 1.$$

---

[3] Thanks to Dave Blackston for pointing out some problems with the original presentation of this material.

Since p is odd, this is a sum of $n + 1$ odd numbers. The only way for that the sum to be odd is if $n + 1$ is odd, which forces n to be even. That means that B is an even power of a prime, so that B must be a perfect square.

Hence, all but one of A, B, C, etc., the factors of N, must be perfect squares, and the other one, suppose it is A, has $\int A$ oddly even.

What about this last factor $\int A$? Euler tells us that if we write $A = q^m$, then for $\int A$ to be oddly even, we must have q a prime number of the form $4n + 1$ and also m must be an odd number of the form $4\lambda + 1$. He doesn't tell us exactly why; probably he planned to add the details later. Perhaps we can fill in the gaps.

Odd prime numbers are either of the form $4n + 1$ or $4n + 3$. First, we will show that our number q cannot be of the form $4n + 3$. Suppose that $q = 4n + 3$. Then powers of q alternate between $+1$ (mod 4) and $+3$ (mod 4) (though Euler wouldn't use the "mod" notation to say this. That notation is due to Gauss in about 1800). Hence,

$$\int A = \int q^m = 1 + q + q^2 + \cdots + q^{m-1} + q^m$$

is either 1 (mod 4) or 0 (mod 4), depending on whether m is even or odd, respectively. It is never 2 (mod 4), so $\int A$ is never oddly even.

On the other hand, if $q = 4n + 1$, then all powers of q are 1 (mod 4). Then

$$\int A = \int q^m = 1 + q + q^2 + \cdots + q^{m-1} + q^m \ (\text{mod } 4)$$

is equal to the number of terms, that is $m + 1$. That is oddly even exactly when m is 1 (mod 4), or, as Euler said, m is of the form $4\lambda + 1$.

Euler summarizes his result saying that, "an odd perfect number will have the form $(4n + 1)^{4\lambda+1} PP$ where P is an odd number and $4n + 1$ is prime."

Since Euler's time, there have been a number of new results about odd perfect numbers, and there is a web page devoted to the subject. [OP] The most striking of these is perhaps that there aren't any that are smaller than 10^{300}. It is remarkable that we know so much about them, but we still don't know if there are any.

References

[Anon] Anon., "Divisor function," Wikipedia, http://en.wikipedia.org/wiki/Divisor_function, October 6, 2006.

[E]   Euclid, *The Thirteen Books of Euclid's Elements*, 3 vols, translated from the text of Heiberg and annotated by Sir Thomas L. Heath, Cambridge University Press, 1908. Reprinted by Dover, New York, in various editions.

[E792] Euler, Leonhard, Tractatus de numerorum doctrina capita sedecim quae supersunt, *Commentationes arithmeticae* 2, 1849, pp. 503–575, reprinted in *Opera Omnia* Series I vol. 5 pp. 182–283. Available through The Euler Archive at www.EulerArchive.org.

[L]   Legendre, Adrien Marie, *Essai sur la théorie des nombres*, Paris, Ches Duprat, an VI [1798].

[G]   Gauss, Carl Friedrich, *Disquisitiones arithmeticae*, translated by Arthur A. Clarke, Yale University Press, New Haven, 1966, revised by W. Waterhouse, et. al., Springer-Verlag, New York, 1986.

[GW]   Greathouse, Charles and Weisstein, Eric W. "Odd Perfect Number." From MathWorld—A Wolfram Web Resource. `http://mathworld.wolfram.com/OddPerfectNumber.html`.

[OP]   `OddPerfect.org`, October 8, 2006. Under development.

[S]   Schooten, Frans van, *Exercitationum mathematicarum libri quinque*, Elsevier, 1656.

[W]   Weil, André, *Number theory: an approach through history from Hamurapi to Legendre*, Birkhäuser, Boston, 1984.

11

Euler and Pell

(April 2005)

Much of a modern course in elementary number theory has its roots in Euler (though the notation is largely due to Gauss.) Euler, in turn, cites as his inspiration the works of Fermat, Diophantus, Goldbach and Pell, among others. This month we will look at the so-called Pell's Equation, $y^2 = ax^2 + 1$, a, x and y integers, named after the English mathematician John Pell (1610–1685), who lived about a hundred years earlier than Euler.

It was Euler who attached the name "Pell's equation" to this formula. People often say that Euler probably made a mistake in attributing the equation to Pell. Boyer [BM] writes

> [Brahmagupta (fl. 628)] suggested also the Diophantine quadratic equation $x^2 = 1 + py^2$, named mistakenly for John Pell (1611–1685) but first appearing in the Archimedian cattle problem. The Pell equation was solved for some cases by Brahmagupta's countryman Bhaskara (1114–ca. 1185)

Pell (whose dates Katz [K] gives as 1610–1685, not quite the same dates as Boyer gives) apparently had nothing to do with it, though Pell was rather secretive, and new evidence may emerge.

Euler's first excursion into Pell's equation was his 1732 paper E29, bearing a title that translates as "On the solution of problems of Diophantus about integer numbers." The main result of this paper is to show how certain quadratic Diophantine equations can be reduced to the Pell equation. In particular, he shows that if we can find a solution to the Diophantine equation $y^2 = an^2 + bn + c$ and we can find solutions to the Pell equation, $q^2 = ap^2 + 1$, then we can use the solutions to the Pell equation to construct more solutions to the original Diophantine equation. He also shows how to use two solutions to a Pell equation to construct more solutions, and notes that solutions to a Pell equation give good rational approximations for $\sqrt{a}$. When Euler discovers the connection between the Pell equation and continued fractions, most of this becomes obsolete, so we will not dwell on it here.

Euler returns to the Pell equation more than 30 years later with his paper "On the use of a new algorithm in solving the Pell problem," E323. The *Summarium* at the beginning of the article announces that the new algorithm will enable us to find easily a solution in the case $a = 61$. This is rather dramatic, since the smallest solution to the equation $p^2 = 61q^2 + 1$ has p a ten-digit number and q a nine-digit one. To find such solutions by hand would indeed be arduous.

Euler begins to describe his algorithm with an example, using $a = 13$. We know that $\sqrt{13}$ is between 3 and 4, so he writes

$$\sqrt{13} = 3 + \frac{1}{a}$$

where we know that $a > 1$. A bit of algebra finds a to be exactly

$$a = \frac{\sqrt{13} + 3}{4}.$$

Knowing that $3 < \sqrt{13} < 4$ makes it easy to show that $1 < a < 2$, so we write

$$a = \frac{\sqrt{13} + 3}{4} = 1 + \frac{1}{b}$$

where, again, $b > 1$. An almost identical calculation shows that

$$b = \frac{4}{\sqrt{13} - 1} = 1 + \frac{1}{c}.$$

If we pause to take stock of what has happened, we get a clue to what Euler is doing here. If we make the substitutions, we get that

$$\sqrt{13} = 3 + \frac{1}{a}$$

$$= 3 + \cfrac{1}{1 + \cfrac{1}{b}}$$

$$= 3 + \cfrac{1}{1 + \cfrac{1}{1 + \cfrac{1}{c}}}$$

Euler is building a continued fraction. He continues, finding, in turn, d, e, and f, and he finds that $f = a$, so the process is cyclic, and, after the initial 3, the coefficients that repeat are 1, 1, 1, 1 and 6. He does some more examples, including $a = 61, 67, 31, 46$, then 54. In each case, he notes that there is an initial integer followed by a pattern that repeats. The initial integer, he denotes v, is the integer part of $\sqrt{a}$. This is followed by a palindromic sequence of integers, followed by the integer $2v$. Then the palindrome and the $2v$ repeat. He gives a table of the cycles for all non-square integers from 2 to 120. He also notes some of the very interesting patterns within these cycles. Hardly any of these properties were in E71, Euler's pioneering paper on continued fractions.

With the existence of these patterns in place, though, he is ready to use some results of his earlier paper. That paper has been translated into English and published in the journal *Mathematical Systems Theory*. [E71] Euler reviews some of his results on how to evaluate continued fractions if you know the pattern of the "indices."

To evaluate the continued fraction corresponding to a sequence of indices, make a table as below:

Indices	v	a	b	c	$\cdots$	m	n	
x	1	v	$av + 1$	$(av + 1)b + v$		M	N	$nN + M$
y	0	1	a	$ab + 1$		P	Q	$nQ + P$

Today we would say that the sequence of numerators, x and of denominators y, each satisfy a recursive relation of order 2, with initial conditions 1, v and 0, 1 respectively.

Let's do an example. The indices for $\sqrt{3}$ are 1, 1, 2, 1, 2, 1, 2, etc., so $v = 1, 2v = 2$, and the palindrome is just 1. To evaluate the first several values of the continued fraction

$$\sqrt{3} = 1 + \cfrac{1}{1 + \cfrac{1}{2 + \cfrac{1}{1 + \cfrac{1}{2 + \text{ etc.}}}}}$$

we start with a table

Indices		1	1	2	1	2	1	2
x	1	v	x_2					
y	0	1	y_2					

Now, $v = 1$ since $1 < \sqrt{3} < 2$. The next index is 1, so $x_2 = 1 \cdot v + 1 = 2$ and $y_2 = 1 \cdot 1 + 0 = 1$, giving

Indices		1	1	2	1	2	1	2
x	1	1	2					
y	0	1	1					

The next index is 2, so we get

Indices		1	1	2	1	2	1	2
x	1	1	2	5				
y	0	1	1	3				

Continuing,

Indices		1	1	2	1	2	1	2
x	1	1	2	5	7	19	26	71
y	0	1	1	3	4	11	15	41

These quotients give progressively better approximations of $\sqrt{3}$, alternating between being too large and being too small. The last one, 71/41, is accurate to three decimal places.

But solutions of the Pell equation $3qq + 1 = pp$ also have quotients that approximate $\sqrt{3}$. In fact, several, but not all of these quotients, give rise to solutions to the equation, $(p, q) = (0, 1), (1, 2), (4, 7),$ and $(15, 26)$. In this particular case, the pattern of solutions and non-solutions is fairly simple, but Euler's paper gives rules for the pattern of solutions in every case. We leave finding and describing these patterns to interested readers and to students in search of a number theory project. Also, some quotients give rise to solutions to $3qq - 1 = pp$, and Euler gives ways to use these solutions to find solutions to Pell's equation.

Eight years later, in 1773, Euler returns to the Pell equation with E559, "New aids for solving the formula $axx + 1 = yy$," not published until 1783. In this paper, he gives ways to generate solutions of the Pell equation from solutions to related equations, $app - 1 = pp, app - 2 = pp, app + 2 = pp$ and $app + 4 = pp$. This provides a kind of converse to the main results of E29, published 50 years earlier.

Euler wrote 96 papers that the editors of the *Opera Omnia* have classified as "number theory." They fill volumes 2, 3, 4 and 5 of series I, about 1700 pages. The three papers we have looked at here comprise only about 2% of Euler's number theory papers, but they extend from one of his very first papers, E29, to one of his last, E559, published in 1783, the year Euler died. Though the Pell equation was a relatively minor aspect of Euler's work, it did hold his interest for his whole life. And it is still interesting today.

References

[BM]   Boyer, Carl B. and Uta C. Merzbach, *A History of Mathematics*, 2nd ed, Wiley, New York,1989.

[K]   Katz, Victor J., *A History of Mathematics*, 2nd ed, Addison-Wesley, Reading, MA, 1998.

[OC]   O'Connor, J. J., and E. F. Robertson, "John Pell," The MacTutor History of Mathematics archive, www-groups.dcs.st-and.ac.uk/~history/Mathematicians/ Pell.html.

[E29]   Euler, Leonhard, De solutione problematum Diophanteorum per numeros integros, *Commentarii academiae scientiarum Petropolitanae*, 6 (1732/3) 1738, pp. 175–188, reprinted in *Opera Omnia* Series I vol. 2 pp. 6–17. Available online at EulerArchive.org.

[E71]   ——, "An Essay on Continued Fractions" translated by Myra F. Wyman and Bostwick F. Wyman, *Math. Systems Theory* 18 (1985) 295–328.

[E323] ——, De solutione problematum Diophanteorum per numeros integros, *Novi commentarii academiae scientiarum Petropolitanae*, 11 (1765) 1767, pp. 28–66, reprinted in *Opera Omnia* Series I vol. 3 pp. 73–111. Available online at EulerArchive.org.

[E559] ——, Nova subsidia pro resolutione formulae $axx + 1 = yy$, *Opuscula analytica* 1, 1783, p. 310–328, reprinted in *Opera Omnia* Series I vol. 4 pp. 76–90.

12

Factors of Forms

(December 2005)

Many number theorists think that the Quadratic Reciprocity Theorem is the most beautiful theorem in all of mathematics. It is said to be Gauss's favorite theorem. Though Euler did not discover quadratic reciprocity, nor did he prove the theorem, he gathered the observational evidence that later guided Legendre and Gauss, so that they would know what to try to prove, and he did manage to prove some preliminary results. In this month's column, we will look at Euler's first results in the subject, results that first appeared in a letter to Christian Goldbach dated August 28, 1742, [JW] and published in 1751 [E164] in the 1744/46 volume of the journal of the St. Petersburg Academy. Harold Edwards wrote about this letter and article in 1983 in a fine paper in *Mathematics Magazine*, [Ed] and he saw the article just a bit differently.

Before we turn to Euler's article, we should remind readers what the Quadratic Reciprocity Theorem tells us. It gives us a way to find when a number a is a perfect square modulo a prime number q. For example, 1 and 4 are perfect squares modulo 7, but, some find it surprising that 2 is also a perfect square, since $3^2 = 9 \equiv 2 \pmod 7$. In the real numbers, $\sqrt{2}$ is irrational, but in the integers modulo 7, $\sqrt{2} = 3$. Just as in the real numbers, nonzero numbers that have square roots have two of them, and the other square root of 2, modulo 7, is 4. The other three nonzero numbers modulo 7 are 3, 5 and 6, and none of them are perfect squares. Euler later proved that modulo any odd prime q, exactly half of the numbers between 0 and p will be perfect squares and half of them will not. The ones that are perfect squares are called *quadratic residues*, and the ones that aren't are called *quadratic non-residues*.

Legendre later introduced a notation to simplify discussions. [M] The so-called *Legendre symbol* is defined for a prime number q and another integer a not divisible by q as follows:

$$\left(\frac{a}{q}\right) = \begin{cases} +1 & \text{if } a \text{ is a perfect square modulo } q \\ -1 & \text{if } a \text{ is not a perfect square modulo } q \end{cases}$$

The symbol is sometimes written as $(a \mid q)$, and is also sometimes defined as being zero if q divides a.

Euler was the first to prove that the product of two quadratic residues or of two quadratic non-residues would be a quadratic residue, but that the product of a residue and a non-residue would be a non-residue. This fact translates into Legendre symbols as

$$\left(\frac{ab}{q}\right) = \left(\frac{a}{q}\right)\left(\frac{b}{q}\right).$$

Because of this fact, we can confine our inquiry to the cases when a is a prime number, since if a is not prime, we can factor a and consider the problem a factor at a time.

The prime number 2 is sometimes a problem case in number theory, being the only even prime number, and often we must deal with it separately. This formula covers the situation:

$$\left(\frac{2}{q}\right) = \begin{cases} +1 & \text{if } q = 1, 7 \ (\text{mod } 8) \\ -1 & \text{if } q = 3, 5 \ (\text{mod } 8) \end{cases}$$

$$= (-1)^{(p^2-1)/8}$$

Euler knew the fact behind this formula, but he apparently never gave a proof of the fact.

The "exponent" notation given in the second line compacts the notation and simplifies calculation, but in my mind it obscures the beauty of the result. Several theorems about quadratic residues have such exponential forms as well as the "modulo" forms I prefer.

We are now ready to state the Quadratic Reciprocity Theorem, which relates the Legendre symbols $\left(\frac{p}{q}\right)$ and $\left(\frac{q}{p}\right)$, as follows:

Quadratic Reciprocity Theorem.   *If p and q are distinct, odd primes, then*

$$\left(\frac{q}{p}\right) = \left(\frac{p}{q}\right)(-1)^{\frac{p-1}{2}\cdot\frac{q-1}{2}}$$

$$= \begin{cases} -\left(\frac{p}{q}\right) & \text{if } p \equiv q \equiv 3 \ (\text{mod } 4) \\ \left(\frac{p}{q}\right) & \text{otherwise} \end{cases}$$

As we mentioned above, apparently Legendre first proved this, and Gauss gave several proofs. Several sources say that Euler stated the theorem in 1783, the year that he died, but nobody seems to give an explicit citation. We will leave that for another column. Here, our purpose is to see how much quadratic reciprocity Euler knew in 1742 when he wrote the letter to Goldbach, and in 1745 when he wrote E164.

Euler's paper *Theoremata circa divisores numerorum in hac forma paa ± qbb contentorum*, "Theorems about divisors of numbers of the form *paa ± qbb*," number 164 in Eneström's index, was only Euler's tenth paper in number theory. He eventually wrote 96 papers in the area, but it is a measure of the relatively low esteem in which number theory was held at the time that half of those papers were only published posthumously.

This particular paper has a very distinctive form, different from any other paper that Euler ever wrote. It does not have the usual "paragraph" structure, but instead is a huge list of 59 "theorems," almost always without proof or discussion, and another 17 "*annotationes*," rather like remarks, all preceded by a single short paragraph. Here is an image of part of the first page of the paper:

✸♨)o(♨✸   -151

THEOREMATA
CIRCA DIVISORES NVMERORVM IN HAC FORMA $paa \pm qbb$ CONTENTORVM.

In fequentibus theorematis litterae a et b defignant numeros quoscunque integros, primos inter fe , feu, qui prae-
ter vnitátem nullum alium' habeant diuiforem communem.

Theorema 1.

Numerorum in hac forma $aa + bb$ coutentorum diui-
fores primi omnes funt vel 2: vel! huius formae $4m + 1$ numeri.

This can be translated as follows:

> In the following theorems, the letters a and b designate arbitrary relatively prime integers, that is, they have only 1 as a common divisor.

> **Theorem 1.** *All of the prime divisors of numbers contained in the form $aa + bb$ are either the number 2 or are numbers of the form $4m + 1$.*

We repeat, Euler gives no proof of Theorem 1 or any of the other theorems in this paper.

After this paragraph and theorem, Euler gives his "theorems" in groups of three. Each triad begins with a theorem giving the forms of prime divisors of the form $aa + pbb$. The second theorem asserts that all of those prime divisors are themselves numbers of this form, and the third theorem is always the contrapositive of the first.

He begins with properties of sums of two square numbers, that is numbers that can be written in the form $aa + bb$. These properties are well known now, and had been noted by Fermat almost a hundred years earlier, but in 1742 number theory was not widely studied, and probably few people other than Euler and Goldbach knew them. His first triad of theorems continues:

> **Theorem 2.** *All prime numbers of the form $4m + 1$ in turn are contained in numbers of this form.*

> **Theorem 3.** *Thus the sum of two squares, that is numbers of the form $aa + bb$ are never divided by any number of the form $4m - 1$.*

Note that Theorem 3 would not be true without the condition in the paragraph at the beginning of the paper that a and b must be relatively prime.

Euler does not mean these as "theorems" in the modern sense of the word. Rather, they are statements he is certain are true, having examined a large number of cases. Almost 20 years later in a paper titled *Demonstratio theorematis Fermatiani omnem numerum primum formae $4n + 1$ esse summam duorum quadratorum*, "Proof of a theorem of Fermat that all prime numbers of the form $4n + 1$ are the sum of two squares." [E241] Euler gives a partial proof of Theorem 1, but he is only able to show that such primes are the sum of squares of two *rational* numbers, not the sum of squares of two *integers*.

He continues with his "theorems" about numbers of the form $aa + 2bb$:

Theorem 4. *The prime divisors of numbers contained in the form $aa + 2bb$ are always either 2 or numbers contained in the form $8m + 1$, or in the form $8m + 3$.*

Theorem 5. *All prime numbers of the forms $8m + 1$ or $8m + 3$ are contained among the numbers of the form $aa + 2bb$.*

Theorem 6. *No number of the form $aa + 2bb$ can be divided by any number of the form $8m - 1$ or of the form $8m - 3$.*

Euler will soon learn a lot more about numbers of the form $aa + 2bb$. In 1753, just two years after this paper is published, he will write another paper, E256, entirely devoted to the properties of such integers. For example, he tells us there, and also gives proofs, that the set of such numbers is closed under multiplication. We will perhaps devote part of a future column to this delightful paper, but there isn't room in this one.

Let us return to E164. Euler continues listing "theorems," three at a time, each describing the divisors of a form $aa + pqq$, for p the prime numbers 3, 5, 7, 11, 13, 17, 19, and then for composite numbers 6, 10, 14, 15, 21, 30 and 35. He demonstrates by example that the theory of forms involving composite values of p is an easy corollary of the theory of forms for which p is prime. A typical example is Theorem 19, giving all the possible forms of prime divisors of a number of the form $aa + 13bb$,

Theorem 19. *All of the prime divisors of a number of the form $aa + 13bb$ are either 2 or 13 or they are described by one of the following 12 formulas*

$$52m + 1 \qquad 52m + 7$$
$$52m + 49 \qquad 52m + 31$$
$$52m + 9 \qquad 52m + 11$$
$$52m + 25 \qquad 52m + 19$$
$$52m + 29 \qquad 52m + 47$$
$$52m + 17 \qquad 52m + 15.$$

This seems like a disorganized jumble of numbers, but there are a great many patterns here. Some of those patterns would be easy to see if we had looked at all 59 of Euler's theorems, but others require Euler's genius to discern, as well as his immense patience and skills at calculation to prepare the data. Today it is a pleasant exercise in Maple™ or Mathematica® to reproduce them. What would Euler have done with such tools?

Here are some of the easier patterns in the cases when p is prime:

- The number 1 is always among the possible remainders.

- There are 12 formulas because p is 13, and 12 is one less than that. In general, the number of formulas necessary to describe the factors of $aa + pbb$ is the number of integers less than p and relatively prime to p.

- The formulas describe numbers modulo 52 because, in general, the possible prime factors are determined by their values modulo $4p$.

- Since we are talking about *prime* factors, the remainders must obviously be relatively prime to 52, and, in general, relatively prime to $4p$. In fact, the possible remainders of prime factors (not counting the two special factors, 2 and p) of numbers of the form $aa + pbb$ are exactly half of the remainders less than $4p$ and relatively prime to $4p$. The other half of the remainders are the ones described in the third theorem of Euler's triads of theorems.

This last pattern leads to what seem to me to be more difficult observations:

• If α is a possible remainder, then $-\alpha$ is always an impossible remainder.

Here, of course, we take the negative modulo $4p$. For example, in the case $p = 13$, we see in Theorem 19 above, we see that 7, 25 and 47 are all among the possible remainders. If we looked at Theorem 21, we would see that their negatives modulo 52, which are the values 45, 27 and 5, respectively, are all among the impossible remainders.

• If α and β are among the possible remainders, then so also is $\alpha\beta$.

We see, for example, that 7 and 11 are possible remainders. Knowing that, modulo 52, their product 77 leaves a remainder 25, we check and see that 25 is also a remainder.

In modern terms, Euler has shown that the set of remainders of prime divisors of numbers of the form $aa + pbb$ modulo $4p$ form a subgroup of index 2 (though he hasn't been explicit about showing that it contains the necessary inverses). Euler, of course, did not have these modern terms. They were at least a hundred years away, and came late enough that Latin had been abandoned as the international language of mathematics. Hence, mathematical Latin does not even have the vocabulary to write these results in the context of group theory.

Finally, we come to the most delicate pattern that Euler found here, and the one that links factors of forms to quadratic reciprocity. If α is relatively prime to $4p$, and also less than $4p$, then the patterns we have already described tell us that either α is among the possible remainders, or $-\alpha$ is, but not both. Euler wants to determine which of these two is the possible remainder.

To begin to explain the pattern he sees, he gives us a table, which we give here, slightly modified:

If		$p = 3n + 1$	then	-3	is a possible remainder. Otherwise, $+3$ is.
If		$p = 5n + 1$			
	or	$p = 5n + 4$	then	$+5$	is a possible remainder. Otherwise, -5 is.
If		$p = 7n + 1$			
	or	$p = 7n + 2$			
	or	$p = 7n + 4$	then	-7	is a possible remainder. Otherwise, $+7$ is.
If		$p = 11n + 1$			
	or	$p = 11n + 3$			
	or	$p = 11n + 4$			
	or	$p = 11n + 5$			
	or	$p = 11n + 9$	then	-11	is a possible remainder. Otherwise, $+11$ is.

These rules all have the same form. We take p to be a prime number, and for another prime number q, we ask whether the possible remainder will be q or $-q$. The pattern here is very subtle. Euler saw the pattern, and only then did he organize his presentation of the data to make it easier to explain the pattern. Even so, it is not very easy.

The possible forms that p might take are each given modulo another prime number q. The remainders in our list are all the perfect squares modulo q. In the last list, for example, where $q = 11$, the numbers 1, 4 and 9 are obviously squares, while 3 and 5 are the squares of 5 and 4, respectively, modulo 11.

The table tells us that, for $q = 5$, $+q$ is a possible remainder if p is a perfect square modulo q. However, for $q = 3$, 7 or 11, exactly the opposite is the case; $-q$ is a possible remainder if p is a perfect square modulo q.

The pattern would be a little less obscure if Euler had extended his table to $q = 13$, so that we could see that the number 13 behaves like the number 5.

So, what property is shared by the primes 5 and 13, but not by the primes 3, 7 and 11? The one that matters is that 5 and 13 are of the form $4m + 1$, but 3, 7 and 11 are of the form $4m + 3$.

Let's try to tie this back to quadratic reciprocity now. Given a prime number p, Euler wants to be able to tell us about the prime divisors of numbers of the form $aa + pbb$. The second theorem in Euler's triads tell us that these are exactly the prime numbers that themselves can be written in this form.

Euler describes these possible prime divisors in terms of their remainders modulo $4p$, and the result that we described as being related to subgroups tells us that we can reduce the problem to remainders that are *prime* remainders.

Then whether q, a prime remainder modulo $4p$ is a possible remainder for a prime of the form $aa + pbb$, depends on whether p is a perfect square modulo q, and on whether q is of the form $4m + 1$. That is:

Euler's Rule. *In the case when q is of the form $4m + 1$, a prime divisor of $aa + pbb$ can have the form $4n + q$ if p is a perfect square modulo q, that is if $\left(\frac{p}{q}\right) = +1$, but not if $\left(\frac{p}{q}\right) = -1$.*

On the other hand, if q is of the form $4m + 3$, the opposite rule applies and it can have the form $4n - q$ if p is a perfect square modulo q, that is if $\left(\frac{p}{q}\right) = +1$, but not if $\left(\frac{p}{q}\right) = -1$.

Perhaps an example would be useful. Let us pretend that we haven't seen Theorem 19, and ask if a number of the form $aa + 13bb$ can have the form $52n + 23$. Here, $p = 13$, $q = 23$, and q is of the form $4m + 3$. So, we look at the second part of Euler's rule and see that a prime divisor can have the form $52n \{23$ if $\left(\frac{p}{q}\right) = +1$, that is if $\left(\frac{13}{23}\right) = +1$. It turns out that $\left(\frac{13}{23}\right) = +1$ (a fact we can determine either by trying all the possibilities and finding that $6^2 = 36 = 13 \pmod{23}$, or by leaving Euler's time for our own and using the Quadratic Reciprocity Theorem). Hence, prime divisors of the form $52n - 23$ are possible, and consequently, those of the form $52n + 23$ are impossible.

When we check Theorem 19, we do not find the number 23 among the possible remainders, so our calculation checks out.

It's not quadratic reciprocity, but it is clearly closely related. Moreover, it is a lesson that the rich and elegant theory of quadratic reciprocity, and its tools and machinery that include modular arithmetic, quadratic forms and class fields have their origins in the ordinary and elementary questions of factoring integers into prime factors.

References

[Ed]   Edwards, Harold M., Euler and Quadratic Reciprocity, *Mathematics Magazine*, vol. 56, No. 5. (Nov., 1983), pp. 285–291. Reprinted in *The Genius of Euler: Reflections on his Life and Work,* edited by William Dunham, Mathematical Association of America, 2007, pp. 233–242.

[E164] Euler, Leonhard, Theoremata circa divisores numerorum in hac forma $paa \pm qbb$ contentorum, *Commentarii academiae scientiarum Petropolitanae* 14 (1744/6), 1751, pp. 151–181, reprinted in *Opera Omnia* Series I vol. 2 pp. 194–222. Available online at `EulerArchive.org`.

[E241] ——, Demonstratio theorematis Fermatiani omnem numerum primum formae $4n + 1$ esse summam duorum quadratorum, *Novi Commentarii academiae scientiarum Petropolitanae* 5, (1754/55) 1760, pp. 3–13, reprinted in *Opera Omnia* Series I vol. 2 pp. 328–337. Available online at `EulerArchive.org`.

[E256] ——, Specimen de usu observationum in mathesi pura, *Novi Commentarii academiae scientiarum Petropolitanae* 6, (1756/57) 1761, pp. 185–230, reprinted in *Opera Omnia* Series 1, vol. 2, pp. 459–492. Available online at `EulerArchive.org`.

[JW]   Juskevic, A. P., and E. Winter, eds., *Leonhard Euler und Christian Goldbach, Briefwechsel 1729–1764*, Akademie-Verlag, Berlin, 1965.

[M]   Miller, Jeff, Earliest Uses of Symbols in Number Theory, `members.aol.com/jeff570/nth.html`, consulted November 24, 2005.

13

2aa + bb
(January 2006)

How do we know what to try to prove?

A logician, or perhaps a Euclidean geometer, might say that we don't *try* to prove anything. We select some axioms or hypotheses. We apply some rules of inference and build a proof. Then the last line of the proof tells us what we've proved.

Uncharitable people sometimes claim that philosophers omit the axioms from this process, that politicians omit the rules of inference, and that people in the humanities never get to the last line.

A scientist, on the other hand, might claim that there is no need for axioms or rules of inference. One need only collect the data, and a correct analysis of the data will reveal the truth.

Much of the culture and image of mathematics is built on this "creation myth," that mathematical theorems are revealed in their statements, and that they are discovered by their proofs. Mathematical truths are imbued with a kind of crystalline purity, true in some absolute sense and unsullied by such vague and uncertain processes like experimentation and creativity.

Today, these notions may seem like idle, post-modernist speculations, but in Euler's time there was a great controversy in science over whether science should be based on observation or on deduction. In rough terms, the sides lined up as Newton vs. Leibniz. Their disagreements weren't based only on the priority dispute in calculus. Leibniz followed in the tradition of Descartes and believed that one should start with known truths and then apply logical methods to discover the truths that must follow from the known truths. Descartes had promoted this basis for reasoning in his *Method*, and used it with great success to discover analytic geometry and to give the first correct explanation of the colors of the rainbow.

Newton, on the other hand, placed a great value on observations. He would make observations, then formulate theories that seemed to explain those observations. He would test those theories, and, if necessary, revise the theories. However, when he explained his theories, he, like Archimedes, would frequently hide the methods by which he made his discoveries. Of course, Newton was not purely "Newtonian," just as Leibniz and Descartes were not purely "Cartesian," but these are the rough outlines of their disputes. Further details are in Hall's fine book *Philosophers at War*. [H]

Euler practiced observation in his work on applied mathematics, though he often hid his method, in the style of Archimedes. He also followed Newton and Descartes in replacing the constructive methods of Geometry with the analytic methods of algebra and calculus.

Early in his career, Euler tended to be Leibnizian and Cartesian. As he matured, he selected principles from both sides of the dispute, but in general he became more and more Newtonian. His exposition, though, seemed more and more Leibnizian, as he developed a very modern-looking style of Theorem - Proof - Corollary. One could think from his writing that he was a "proof machine" that never made an observation or made a conjecture.

In 1756, Euler decided to "come clean" about how he knew what to try to prove. He wrote a paper, *Specimen de usu obserationum in mathesi pura*, "Example of the use of observation in pure mathematics," [E256] in which he describes his path from observation to theorem. He attributes the technique, on slim evidence, to Fermat.

To explain his method, Euler selects material from some of his then-recent papers on number theory, especially E164 (the principal subject of the December 2005 column "Factors of forms"), about the quadratic forms $aa + pbb$, and E241, in which he gives his proof that the prime numbers of the form $4n + 1$ are exactly the ones that are the sum of two squares.

Here in E256, Euler studies numbers of the form $2aa + bb$, a special case of the numbers he studied in E164. After a two page introduction about the relation between observation and proof, Euler begins his work with eight observations about numbers of the form $2aa + bb$, taking a and b to be relatively prime. This takes only two pages. He spends the last 15 pages of the paper trying to prove these eight observations. As it turns out, he isn't able to prove all of them, and the things that were hardest to observe aren't always the hardest ones to prove. In the course of his proofs, though, Euler comes across other things that are true, and proves them, too.

Euler begins his observation with a list. He tabulates all the numbers less than 500 of the form $2aa + bb$, with a and b relatively prime. His list looks something like this:

$2 + bb$)   3, 6, 11, 18, 27, 38, 51, 66, 83, 102, 123, 146, 171, 198, 227, 258, 291, 326, 363, 402, 443, 486.

$8 + bb$)   9, 17, 33, 57, 89, 129, 177, 233, 297, 369, 449.

$\cdots$

$450 + bb$)   451, 454, 466, 499.

Note that Euler expects both a and b to be nonzero. Now he starts mining his list for information and making his observations:

Observation 1.   We look at the 45 prime numbers that appear on the list:

$$3, \ 11, \ 17, \ 19, \ 41, \ 43, \ 59, \ 67, \ 73, \ldots, \ 491, \ 499.$$

None of these numbers appears more than once on the list, hence we speculate that such prime numbers are *uniquely* represented in this form. As Euler gets to proving the theorems behind these observations, it becomes clear that this is really two statements, first that prime numbers appear only once, and second that (odd) numbers that appear only once are prime.

Observation 2.   Next we list the doubles of the prime numbers.

$$6, \ 22, \ 34, \ 38, \ 82, \ 86, \ 118, \ 134, \ 146, \ldots, \ 466, \ 482.$$

They, too, appear only once each, so they, too, are uniquely represented. They are exactly the doubles of the primes in the first list, and there are no numbers on the list that are multiples of 4.

Since before Euclid's time, about 2400 years ago, numbers that are divisible by 2 but not divisible by 4 have been called "oddly even." Observation 2 says, among other things, that even numbers of the form $2aa + bb$ are oddly even. Knowing this, we quote Euler's next observation.

Observation 3.   "Compare the numbers that are odd and the ones that are even, but oddly even, and I observe: *If an odd number is represented, then so also is its double, and also, if an even number appears, half of it will appear as well.*"

Observation 4.   For those remaining numbers (i.e., not prime, also not even) list their prime factorization, and at the same time, in parentheses give the number of times each number appears in the list:

$$3^2(1)3^3(1)3 \cdot 11(2), \ 3 \cdot 17(2)3 \cdot 19(2), \ 3^4(1), \ 3^2 \cdot 11(2), \ 11^2(1), \ 3 \cdot 41(2), \text{ etc.}$$

From this we see that any product of the prime numbers we saw in Observation 1 also occurs on the list, and it occurs more than once if it is composed of different factors. For example, 33 has two prime factors, 3 and 11, and it occurs twice on the list because it has two different representations of the form $2aa + bb$, being $2 \cdot 4 + 25$ and $2 \cdot 16 + 1$.

Note that Euler does NOT claim, though it is true, that the number of times a number occurs doubles for each odd prime factor it has.

To deal with the special prime number 2, Euler specifically notes that we can get it by taking $b = 0$ and $a = 1$, despite his general assumption that both a and b be nonzero.

Observation 5.   Among the factors of these numbers there are no primes except those that are also of the form $2aa + bb$.

Observation 6.   No prime numbers of the forms $8n - 1$ or $8n - 3$ are of the form $2aa + bb$, nor can they be divisors of numbers of the form $2aa + bb$, as long as a and b are relatively prime.

Observation 7.   No number of the form $2aa + bb$, with a and b relatively prime has any prime divisors other than 2 and prime numbers of one or the other forms $8n + 1$ or $8n + 3$.

Observation 8. Now it is of greatest interest to observe that every prime number of these two forms $8n + 1$ and $8n + 3$ occurs on the list.

Euler notes that all of these observations are easy to make, and some of them can be proved, but for others the proofs are "most difficult." Into the first category (the easy ones) fall observations 1, 2, 3, 4, and the first part of 6. The hard ones are 5, the second part of 6, and 7. The very deepest, he says, is Observation 8. Moreover, he notes, these properties are similar in many ways to the properties of the sums of two squares that he described in E228 and E241.

Now that he has shown us how observations give him ideas about what to prove, the character of this paper changes dramatically. Euler sets out to prove the things he observed, and he uses his usual Theorem-Proof-Corollary structure, with a few examples put in near the end.

As we mentioned above, Euler does not prove the same things he observes, and when he proves these observations, he doesn't prove them in the same order he observed them, either. His first theorem is a proof of the first part of Observation 3. We quote:

Theorem 1. *If N is of the form $2aa + bb$, then so is its double.*

Proof. Take $N = 2mm + nn$, so that $2N = 4mm + 2nn$. Take $2m = k$. This makes $2N = kk + 2nn$, and so $2N$ is a number of the form $2aa + bb$. Q.E.D.

Predictably, Theorem 2 is the converse of Theorem 1, and completes his proof of Observation 3, which, in turn, implies the result in Observation 2. Again we quote:

Theorem 2. *If a number $2N$ is of the form $2aa + bb$, then so also its half, N is of the same form.*

Proof. For this to happen, it is necessary that nn be even, and so n itself is even. Write $n = 2k$, so that $2N = 2mm + 4kk$ and so $N = mm + 2kk$, which is a number of the form $2aa + bb$.   Q.E.D.

Something about this gave Euler the idea of asking whether the product of two numbers of the form $2aa + bb$ was again a number of that form, even though he had not made any observations about the question. He answers the question in the affirmative, and adds a little bit, with Theorem 3. Euler's proof of his Theorem 3 is a bit longer and wordier, so we only paraphrase it:

Theorem 3. *If M and N are of the form, then so is their product MN.*

Proof. Take $M = 2aa + bb$ and $N = 2cc + dd$. Then

$$MN = 4aacc + 2aadd + 2ccbb + bbdd$$
$$= 4aacc + 4acbd + bbdd + 2aadd - 4acbd + 2ccbb$$
$$= (2ac + bd)^2 + 2(ad - cb)^2$$

Note also that if we reverse the signs of the added terms, we get MN in a second way, as

$$MN = 4aacc + 2aadd + 2ccbb + bbdd$$
$$= 4aacc - 4acbd + bbdd + 2aadd + 4acbd + 2ccbb$$
$$= (2ac - bd)^2 + 2(ad + cb)^2$$

Moreover, these two ways must be different.

Note that there might be a little hang-up if any of the calculated values come out negative. If that happens, take their corresponding positive values.   Q.E.D.

Surely Euler noticed that this same proof would work for any quadratic form $paa+bb$, but he does not mention that here. He is doing a good job maintaining his focus on the particular form $2aa + bb$.

Now Euler turns to his first observation, that the prime numbers appear only once on the list. In his Theorem 4, he states the result in its contrapositive, and then proves it by contradiction. Although the proof is a bit long, it is a very clever proof. Also, Euler doesn't do such straightforward proofs by contradiction very often, so it is interesting for that as well. Again, we paraphrase:

Theorem 4. *Any number that can be resolved in two ways into a form $2aa + bb$ is not prime.*

Proof. Euler uses proof by contradiction.

Suppose N is prime and N can be resolved in two different ways. Say $N = 2aa + bb$ and $N = 2cc + dd$, with a and b different from c and d. Multiply the first resolution by cc and the second by aa and subtract to get, on the one hand, $(aa-cc)N$, and on the other hand $aadd - bbcc$, which factors as the difference of squares as $(ad - bc)(ad + bc)$.

Since N is prime, it must divide one or the other of these factors. This is the consequence that Euler will contradict.

Now also, add the two forms and get

$$2N = 2aa + bb + 2cc + dd.$$

From this take away $2ad + 2bc$ and there remains

$$2N - 2ad - 2bc = 2aa + bb + 2cc + dd - 2ad - 2bc$$

which can be reorganized as

$$2N - 2ad - 2bc = aa + (a - d)^2 + cc + (c - b)^2.$$

Now, the RHS is the sum of four squares, and so it is certainly greater than zero, therefore

$$2N - 2ad - 2bc > 0, \quad \text{and}$$
$$N > ad + bc$$

N must also be greater than $ad - bc$, and so N can divide neither $ad + bc$ nor $ad - bc$—the consequence we flagged above cannot be true.

All this followed from the hypothesis that there were two resolutions, so there can't be two distinct resolutions of a prime number.   QED.

Observation 5 is next on Euler's agenda, that all prime factors of a number of the form $2aa + bb$ (where a and b are relatively prime) must also be of that form. This is the first of the observations that Euler described as being more difficult. Indeed, his Theorems 5 to 8 are rather difficult technical lemmas that lead to Theorem 9, which we quote:

Theorem 9. *No number of the form $2aa + bb$, for which a and b are relatively prime, can have a prime factor that is not also of this form.*

Theorems 10 and 11 explain Observation 1 and the first part of Observation 6, respectively:

Theorem 10.   *If a number of the form* $2aa + bb$ *resolves into this form in just one way, and if a and b are relatively prime, then the number is certainly prime.*

Theorem 11.   *No number of the forms* $8n - 1$ *or* $8n - 3$ *can divide any number* $2aa + bb$, *as long as a and b are relatively prime.*

Euler's last theorem in this vein is related to the other results, but it is not explicit among his eight observations:

Theorem 12.   *If a number in one or the other of the forms* $8n + 1$ *or* $8n + 3$ *cannot be resolved into the form* $2aa + bb$, *then it is not prime; and if it can be so resolved in exactly one form, then it is prime; and if it can be resolved in more than one such way, then it is not prime, but it is composite.*

Except for the part about observations at the beginning, this paper really has turned into a fairly typical Euler paper in number theory. In true Eulerian form, there are two examples.

First, Euler asks whether the number 67579 is prime. He sees that it is of the form $8n + 3$, and so, by Theorem 12, he can show it is prime by showing that it is uniquely of the form $2aa + bb$. He does this more or less by brute force, and finds that $67579 = 2 \cdot 87^2 + 229^2$, and this is its only representation in the form $2aa + bb$, hence it is prime.

In Euler's time, 67579 is not a particularly large prime number. They knew several seven-digit prime numbers, but didn't know any eight-digit ones yet.

Euler's second example is to demonstrate that 40081 is not a prime number. Though it is of the form $8n + 1$, it is not of the form $2aa + bb$, so, again by Theorem 12, it is not prime. In fact, it is the product of 149 and 269, but nothing in this technique helps to find the factorization.

If Euler had ended the paper here, the bulk of this paper would be much like many of his other papers, but he has a surprise ending, two entirely unexpected theorems about square and triangular numbers that are corollaries of the theorems he has already proved:

Theorem 13.   *If a number n is in no way the sum of a square and a triangular number, then the number* $8n + 1$ *certainly is not prime.*

Proof.   If n were not of the form $aa + \frac{1}{2}(bb + b)$, then $8n + 1$ could not be of the form

$$8aa + 4bb + 4b + 1,$$

and hence could not be (taking $p = 2a$ and $q = 2b + 1$) of the form $2pp + qq$, and so could not be prime.   Q.E.D.

This is a negative result. Its contrapositive, that if $8n + 1$ is prime, then n is a square plus a triangle, is only a necessary, and not a sufficient condition. For (my) example, $10 = 9 + 1$ is a square and a triangle, but 81 is not prime.

Theorem 14.   *If n is in no way the sum of a triangular number and the double of a triangular number, then* $8n + 3$ *is certainly not prime.*

Proof. Here our number does not start as $aa + a + \frac{1}{2}(bb + b)$, which, multiplying by 8 and adding 3 (in the form $2 + 1$) gives $8aa + 8a + 2 + 4bb + 4b + 1$, which is of the form $2pp + qq$. Hence, our number is *not* of this form, and cannot be prime. Q.E.D.

The Reader is encouraged to verify these surprising results with some further experiments.

This ends Euler's remarkable description of the delicate dance between observation and deductive proof, complete with examples and a surprise ending.

References

[E164] Euler, Leonhard, Theoremata circa divisores numerorum in hac forma $paa \pm qbb$ contentorum, *Commentarii academiae scientiarum Petropolitanae* 14 (1744/6), 1751, pp. 151–181, reprinted in *Opera Omnia* Series I vol. 2 pp. 194–222. Available online at EulerArchive.org.

[E228] ——, De numeris, qui sunt aggregate duorum quadratorum, *Novi Commentarii academiae scientiarum Petropolitanae* 4, (1752/53) 1758, pp. 3–40, reprinted in *Opera Omnia* Series I vol. 2 pp. 295–327. Available online at EulerArchive.org.

[E241] ——, Demonstratio theorematis Fermatiani omnem numerum primum formae $4n + 1$ esse summam duorum quadratorum, *Novi Commentarii academiae scientiarum Petropolitanae* 5, (1754/55) 1760, pp. 3–13, reprinted in *Opera Omnia* Series I vol. 2 pp. 328–337. Available online at EulerArchive.org.

[E256] ——, Specimen de usu observationum in mathesi pura, *Novi Commentarii academiae scientiarum Petropolitanae* 6, (1756/57) 1761, pp. 185–230, reprinted in *Opera Omnia* Series I, vol. 2 pp. 459–492. Available online at EulerArchive.org.

[H] Hall, A. R., *Philosophers at War: The Quarrel between Newton and Leibniz*, Cambridge University Press, 1980.

Part III

Combinatorics

14

Philip Naudé's Problem

(October 2005)

In the days before email, mathematics journals, the MathFest, MAA Online and annual Joint Mathematics Meetings, mathematicians had far fewer ways to communicate. Finished works would appear in books and in a few general scientific and scholarly journals like *Acta Eruditorum* or the *Mémoires de l'Académie des Sciences de Berlin*. Euler usually published in the journal of the St. Petersburg Academy, *Commentarii academiae scientiarum Petropolitanae*, or one of its successor journals.

There are other hidden and informal means of scientific communications that are essential to the health of mathematics. These are the ways we answer questions like "What are you working on?" or "What have you discovered that you haven't published yet?" or "What would be interesting or useful to work on next?" Within a single department, we get to ask and sometimes answer these questions personally. To ask them in a larger community, we have email and meetings.

They couldn't do that in Euler's day. Occasionally, someone would go on a "Grand Tour" and visit scientists in several places. We are only sure that Euler did this once, when he stopped to visit Christian Wolff in the 1720s while moving from Basel to St. Petersburg, though he may have made some shorter trips within Germany in the 1740s and 1750s. Many more people traveled to see Euler, but it wasn't enough to sustain a scientific conversation.

Instead, they wrote letters. Over a thousand letters written by Euler survive, and, because he was a very well organized person, over two thousand letters addressed to Euler still exist. This is why so many of Euler's ideas can be traced to his correspondence. His interest in number theory, for example, began in letters from Goldbach in the 1720s. Other ideas first appear in letters to various Bernoullis, to Stirling, or to Lagrange. Only about a thousand of these letters have been published, though a complete index is in the *Opera Omnia*, Series IV-A, vol. 1. Every once in a while, previously unknown letters turn up and the list grows a little bit.

This month's column has its origins in one of those letters. On September 4, 1740, Philip Naudé the younger (1684–1747) wrote Euler from Berlin to ask "how many ways can the number 50 be written as a sum of seven different positive integers?"

The number 5, for example, can be written as the sum of two different positive integers in exactly two ways, $1+4$ and $2+3$. Naudé wanted the corresponding number of partitions of 50 into seven distinct parts.

The problem seems to have captured Euler's imagination. Euler gave his first answer on April 6, 1741, in a paper he read at the weekly meeting of the St. Petersburg Academy. That paper was published ten years later and is number 158 on Eneström's index. Euler solved the problem in a different way in the *Introductio in analysin infinitorum*, E101, published in 1748, and made more improvements in a paper *De partitione numerorum*, "On the partition of numbers," E191, written in 1750, published in 1753. Late in his life, in 1769, he returned to the problem in *De partitione numerorum in partes tam numero quam specie datas*, "On the partition of numbers into a given kind or number of parts," E394, published in 1770. Readers at the time might have been a bit confused by the order of the publications. The first one written was the second one published.

All three of the articles, E158, E191 and E394, are reprinted in the *Opera Omnia* volumes on Number Theory, though we will see that they could just as well have been classified as papers on Series and put in volumes 14 or 15, or classified as papers on combinatorics and published in volume 7 "Pertaining to the theory of combinations and probability."

The solutions given in E158 and E191 are both based on the relations comparing a series, $a + b + c+$ etc., the sum of its powers, $a^n + b^n + c^n+$ etc., and the sum of products of its terms taken without repeated factors, first two at a time, $ab + ac + bc + ad + bd + cd+$ etc., then three at a time, $abc + abd + acd + bcd + abe+$ etc.

The solutions in E394 and in the *Introductio*, E101, are based on more familiar ideas of multiplying polynomials. We will describe the version given in Chapter XVI of the *Introductio*, "On the Partition of Numbers."

To prepare ourselves for Euler's ideas, let's do a simple example. Let's ask, what is the coefficient of x^5 if we expand

$$(1 + x + x^3)(1 + x^2 + x^3 + x^4)(x + x^2)?$$

We could answer the question by brute force by finding the ninth degree polynomial that results. On the other hand, since we are only asked to find one of the ten coefficients, we can use the addition properties of exponents, and ask how the exponent 5 can arise. We get a product of degree 5 by adding an exponent from the first factor, 0, 1 or 3, to an exponent from the second factor, 0, 2, 3 or 4, and then adding an exponent from the third factor, 1 or 2, and having the resulting sum be 5. The coefficient of x^5 will be the number of ways of forming this sum. Since there are five ways to do this, $(0, 3, 2)$, $(0, 4, 1)$, $(1, 2, 2)$, $(1, 3, 1)$ and $(3, 0, 2)$, the coefficient of x^5 must be 5.

Conversely, since the product expands to

$$x + 2x^2 + 2x^3 + 4x^4 + 5x^5 + 4x^6 + 3x^7 + 2x^8 + x^9,$$

and since the coefficient of x^5 is 5, there are 5 ways to form the desired sum. This is Euler's basic idea, to use the coefficients of polynomials to count things. It is a remarkably powerful idea, as we shall see. Euler begins by comparing a product of polynomials in two variables, possibly as an infinite product, with its expansion into a series. He supposes that

$$(1 + x^\alpha z)(1 + x^\beta z)(1 + x^\gamma z)(1 + x^\delta z) \cdots = 1 + Pz + Qz^2 + Rz^3 + Sz^4 + \cdots$$

and asks for the values of P, Q, R, etc., in terms of α, β, γ, etc.

He answers his own question and explains that P is the sum of powers $x^\alpha + x^\beta + x^\gamma + x^\delta +$ etc. Then Q is the sum of the products of powers taken two at a time, R those taken three at a time, etc.

There may be coefficients, though. In P, the coefficient on x^n tells us how many times the number n occurs in the sequence α, β, γ, etc. In Q, the coefficient tells us how many different ways n can be formed as the sum of a pair of numbers from the sequence. In R, it tells us how many ways it is the sum of three numbers from the sequence, and so forth.

Got it? Maybe not. Let's do an example. Euler writes, in Chapter 16 of Book 1 of his *Introductio* [E101], "In order that this may become clearer, let us consider the following infinite product."

$$(1 + xz)(1 + x^2 z)(1 + x^3 z)(1 + x^4 z)(1 + x^5 z) \cdots.$$

Euler bravely expands this to get

$$
\begin{aligned}
1 &+ z(x + x^2 + x^3 + x^4 + x^5 + x^6 + x^7 + x^8 + x^9 + x^{10} + \cdots) \\
&+ z^2(x^3 + x^4 + 2x^5 + 2x^6 + 3x^7 + 3x^8 + 4x^9 + \cdots) \\
&+ z^3(x^6 + x^7 + 2x^8 + 3x^9 + 4x^{10} + 5x^{11} + 7x^{12} + \cdots) \\
&+ z^4(x^{10} + x^{11} + 2x^{12} + 3x^{13} + 5x^{14} + 6x^{15} + 9x^{16} + \cdots) \\
&+ z^5(\cdots) + \cdots
\end{aligned}
$$

Euler carries the expansion up through z^8 and gives about ten terms of x in each coefficient of z^n. It is quite impressive.

This series is a wealth of information. For example, there is a term $6z^4 x^{15}$. The coefficient 6 tells us that there are six ways to write the exponent of x, 15, as the sum of 4, the exponent of z, different numbers from the sequence 1, 2, 3, 4, 5, ... And, indeed, the six ways are $(1, 2, 3, 9)$, $(1, 2, 4, 8)$, $(1, 2, 5, 7)$, $(1, 3, 4, 7)$, $(1, 3, 5, 6)$ and $(2, 3, 4, 6)$. Similarly, there is a term $15z^7 x^{35}$ (Euler includes it, but we didn't). The coefficient 15 tells us that there are 15 ways to write the number 35 as the sum of seven different positive integers.

Recall that Philip Naudé had asked the number of ways to write 50 as the sum of seven different positive integers, so we need only find the coefficient of $z^7 x^{50}$. He doesn't mention this in the *Introductio*, though he tells us in E158 that the required coefficient is 522, and that this is "a most perfect solution to the problem of Naudé."

There is even more information in the series. If we take $z = 1$ and combine like terms, then the coefficient of x^n will be the number of ways to write n as the sum of one distinct positive integer, plus the number of ways to write it as the sum of two of them, plus the number of ways to write it as the sum of three of them, etc. To say this more directly, the coefficient of x^n is the number of ways to write n as the sum of distinct positive integers.

Of course, Euler does an example. If $z = 1$, then the infinite product is

$$(1 + x)(1 + x^2)(1 + x^3)(1 + x^4)(1 + x^5) \text{ etc.}$$

This expands to give

$$1 + x + x^2 + 2x^3 + 2x^4 + 3x^5 + 4x^6 + 5x^7 + 6x^8 + \text{ etc.}$$

Euler points out that the term $6x^8$ tells us that there are six ways to write eight as the sum of distinct positive integers, and he lists them:

$$8 = 8 \qquad 8 = 7 + 1 \qquad 8 = 6 + 2$$
$$8 = 5 + 3 \qquad 8 = 5 + 2 + 1 \qquad 8 = 4 + 3 + 1$$

He takes pains to note that we must count the "sum" $8 = 8$. He means us to know without being told that $8 = 7 + 1$ is counted as the same sum as $8 = 1 + 7$.

Euler is nowhere near exhausting the idea of using coefficients to count things. He asks, what if we relax the requirement that our summands be distinct, and allow sums like $8 = 3 + 3 + 2$? He tells us that "The condition that the numbers be different is eliminated if the product is put into the denominator." This is a bit cryptic. To clarify it a bit, he asks us to consider the expression

$$\frac{1}{(1 - x^\alpha z)(1 - x^\beta z)(1 - x^\gamma z)(1 - x^\delta z) \cdots}$$

The modern mathematician might ask whether or not this expression converges, but as usual, Euler trusts his notation, and doesn't worry about such things.

The quotient can be expanded into a product of geometric series, which can then be expanded again into a series similar to the one he examined earlier. Euler gives copious details and a good deal of generality, but the case that is most interesting is when α, β, γ, δ are 1, 2, 3, 4, etc. (The second most interesting case is when they are 1, 3, 5, 7, etc.) Then the quotient

$$\frac{1}{1 - x^\alpha z} = 1 + xz + x^2 z^2 + x^3 z^3 + x^4 z^4 + \text{ etc.}$$

The next quotient,

$$\frac{1}{1 - x^\beta z} = 1 + x^2 z + x^4 z^2 + x^6 z^3 + x^8 z^4 + \text{ etc.}$$

Similarly for the infinitely many other quotients. Again, Euler bravely expands the expression

$$\frac{1}{(1 - zx)(1 - zx^2)(1 - zx^3)(1 - zx^4) \text{ etc.}}$$

into a series

$$1 + z(x + x^2 + x^3 + x^4 + x^5 + x^6 + x^7 + \cdots)$$
$$+ z^2(x^2 + x^3 + 2x^4 + 2x^5 + 3x^6 + 3x^7 + 4x^8 + \cdots)$$
$$+ z^3(x^3 + x^4 + 2x^5 + 3x^6 + 4x^7 + 5x^8 + 7x^9 + \cdots)$$
$$+ z^4(x^4 + x^5 + 2x^6 + 3x^7 + 5x^8 + 6x^9 + 9x^{10} \cdots)$$
$$+ z^5(\cdots) + \text{ etc.}$$

The coefficients here tell us how many different ways the exponent of x can be written as the sum of the number in the exponent of z positive integers. Euler gives the example, 13 can be written as the sum of five positive integers (not necessarily *different* positive integers) in 18 different ways because the coefficient of $x^{13}z^5$ turns out to be 18. Euler does not list the 18 different ways.

Now if we take $z = 1$, as before, we get

$$\frac{1}{(1-x)(1-x^2)(1-x^3)(1-x^4)\text{ etc.}} = 1+x+2x^2+3x^3+5x^4+7x^5+11x^6+15x^7+\text{ etc.}$$

The coefficients of this series are called the partition numbers, and give the number of ways the corresponding exponent can be written as the sum of any number of positive integers. For example, the term $11x^6$ tells us that 6 can be written as a sum in 11 different ways. Euler lists them for us.

From here, paths lead in many directions. We could look at Euler's work on the partition numbers themselves. He discovered an amazing recursive relation involving the pentagonal numbers. This line of study continued through Ramanujan, who discovered a number of beautiful and astonishing properties.

Instead of dwelling on this, we will only sketch a few other results and mention a bit about where they lead.

One thing Euler does, but we won't give details, is to show that the coefficients of the expansion

$$\frac{x^3}{(1-x)(1-x^2)} = x^3 + x^4 + 2x^5 + 2x^6 + 3x^7 + 3x^8 + 4x^9 + \text{ etc.}$$

give the number of ways that a number can be written as the sum of two different numbers. For example, $4x^9$ tells us that there are four ways to write 9 as the sum of two different numbers. They are $1+8$, $2+7$, $3+6$ and $4+5$. Likewise, $3x^8$ corresponds to the sums $1+7$, $2+6$ and $3+5$.

Then he shows that the coefficient of x^n in the expansion also gives the number of ways that $n-3$ can be given as the sum of 1's and 2's. Here, $3x^8$ tells us that there will be three ways to write $8-3 = 5$ as a sum of 1's and 2's. They are $1+1+1+1+1$, $1+1+1+2$ and $1+2+2$. Note that it is not necessary to include both 1's and 2's, so $1+1+1+1+1$ is an admissible sum.

Euler also showed how to find similar relations between the number of sums of k different numbers and the number of sums involving only the numbers 1 through k. He did all of his analysis algebraically.

We could also explore the other ways that Euler used the tools of generating functions that he first developed in his study of partitions in E158.

In 1878, J. J. Sylvester, then teaching at Johns Hopkins, and his student Fabian Franklin, working with a group of other graduate students at Hopkins, gave a wonderful graphical proof of the same result, as well as a number of new results. Their paper was based on the insight that arrangements of 16 stars like the following,

$$* \ * \ * \ * \ * \ *$$
$$* \ * \ * \ * \ * \ *$$
$$* \ * \ * \ *$$

can be viewed as giving the number 16 as a sum of three numbers, $6 + 6 + 4$, or as a sum involving only numbers 1 through 3, here $3 + 3 + 3 + 3 + 2 + 2$.

Rather than follow these paths now, we will leave them for another column.

References

[E101] Euler, Leonhard, *Introductio in analysin infinitorum*, 2 vols., Bousquet, Lausanne, 1748, reprinted in the *Opera Omnia*, Series I vols. 8 and 9. English translation by John Blanton, Springer-Verlag, 1988 and 1990. Facsimile edition by Anastaltique, Brussels, 1967.

[E158] ——, Observationes analyticae variae decombinationibus, *Commentarii academiae scientiarum Petropolitanae* 13 (1741/43) 1751, pp. 64–93, reprinted in *Opera Omnia* Series I vol. 2 pp. 163–193. Available online at EulerArchive.org.

[E191] ——, De partitione numerorum, *Novi Commentarii academiae scientiarum Petropolitanae* 3 (1750/51) 1753, pp. 125–169, reprinted in *Opera Omnia* Series I vol. 2 pp. 254–294. Available online at EulerArchive.org.

[E394] ——, De partitione numerorum in partes tam numero quam specie datas, *Novi commentarii academiae scientiarum Petropolitanae* 14 (1769): I, 1770, pp. 168–187, reprinted in *Opera Omnia* Series I vol. 3, pp. 131–147. Available online at EulerArchive.org.

[SF] Sylvester, J. J. and Fabian Franklin, A Constructive Theory of Partitions, Arranged in Three Acts, an Interact and an Exodion, *American Journal of Mathematics*, 5 No. 1 (1882), pp. 251–330.

15

Venn Diagrams

(January 2004)

Almost all of us have seen and used Venn Diagrams. Though they are no longer quite as ubiquitous as they were during the heyday of "The New Math" (has it already been 35 years?), they are still a staple of Discrete Mathematics courses. Nobody calls them anything except "Venn Diagrams," and several modern sources (e. g. [R]) refer to John Venn's 1880 article [V], so it is natural to assume that Venn discovered them. That would be the end of it, unless we read Venn's article (fairly easily available via InterLibrary Loan) and see that Venn doesn't call them "Venn Diagrams." He calls them "Eulerian Circles"! We quote his opening sentences:

> Schemes of diagrammatic representation have been so familiarly introduced into logical treatises during the last century or so, that many readers, even those who have made no professional study of logic, may be supposed to be acquainted with the general nature and object of such devices. Of these schemes one only, viz. that commonly called "Eulerian circles," has met with any general acceptance....

The "Schemes of diagrammatic representation" Venn writes about are ways to represent propositions by diagrams. Euler had written about them almost 120 years earlier in his *Letters of Euler on Different Subjects in Natural Philosophy*, also known as *Letters to a German Princess*. Euler wrote these *Letters*, approximately two per week, from April 1760 to May 1763, as lessons in elementary science for the Princess of Anhalt Dessau. They fill two volumes, over 800 pages, and cover a huge range of topics in science: light, color, gravity, astronomy, mechanics, sound, music, heat, weather, logic, magnetism, optics, and more. Though she was a German princess, the *Letters* were written in French, the language of courtly society. They were published in 1768, in part through the efforts of the French mathematicians Condorcet and Lacroix, and immediately became very popular. Soon, they were translated into all the major languages of Europe. Bill Dunham writes "*Letters to a German Princess* remains to this day one of history's finest examples of popular science." I am working from two editions, 1823 and 1833, of a 1795 English translation by Henry Hunter.

Euler uses his circles in four letters, CII to CV of Volume I, written between February 14 and February 24, 1761. Each letter has a title. CII, for example, is called "On the Perfections of a Language. Judgments and Nature of Propositions, affirmative and negative; universal or particular." The other titles are a little shorter, and mention Forms, Syllogisms and Propositions.

Syllogisms come in a variety of forms. One form is called the *affirmative particular*: Some *A* is *B*. Euler illustrates this with what is labeled "Fig. 45." Euler notices, though, that the figure does not have a unique interpretation. He offers four different interpretations of Figure 45:

I.   Some *A* is *B*.

II.   Some *B* is *A*.

III. Some *A* is not *B*.

IV.   Some *B* is not *A*.

There are more than fifty such figures. Euler uses Figure 27 to justify the following logical argument:

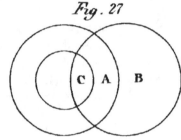

1.   Some *C* is *B*.

2.   All *C* is *A*.

3.   Therefore, some *A* is *B*.

Euler is not thinking of sets. He is using these diagrams to explain syllogisms, propositions and logical arguments. Note also that it doesn't seem to occur to Euler that some of the regions shown in his diagram may be empty.

A hundred and twenty years later, John Venn returns to the same problem, how to use diagrams to represent language. Again, he is not yet interested in sets, but in logic, statements and propositions. Just as Euler notes that several statements correspond to Figure 45, Venn notes that several diagrams may correspond to the same statement. For example, he gives the three diagrams below that correspond to the statement "Some *X* is not *Z*."

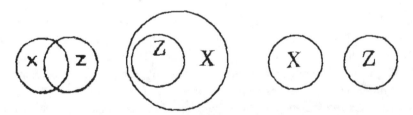

So far, it seems as if Venn is just re-writing Euler using letters from the end of the alphabet instead of letters from the beginning. Some people have suggested that Venn was nothing but an arrogant usurper, taking Euler's ideas and promoting them as his own. Indeed, if Venn stops here, he probably is guilty of that. Let's look at the rest of the paper.

Venn notices that the simple two-circle diagram shown at
the right divides the plane into four regions. He uses bars to
denote "not", so the expression $X\overline{Y}$ denotes "X but not Y",
and is represented by the crescent-shaped region at the left of
the diagram. He continues:

Now conceive that we have to reckon also with the presence, and consequently
with the absence, of Z. We just draw a third circle intersecting the two above,
thus,

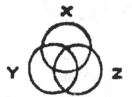

and we have the eight compartments or classes which we need. The subdivisions
thus produced correspond precisely with the letter-combinations.

Venn continues with this idea. He sees that he can not draw diagrams for four propositions using circles, but he can do it using ellipses. The diagram on the left below shows
how this can be done. The compartment marked with an "x" corresponds to the class
$XYZ\overline{W}$.

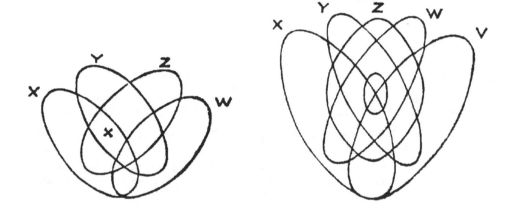

Venn isn't able to find an arrangement of circles or ellipses that diagrams five propositions, though he does find the remarkable one shown above on the right. The regions for
V, W, X and Y are ellipses, and the region for Z is shaped like a donut. People are still
interested in designing Venn diagrams that represent large numbers of propositions. Frank
Ruskey [R] shows a single shape that creates a Venn diagram for seven propositions.

Venn has a key idea that Euler overlooks; he can use the same diagram to analyze
different lists of propositions if he uses them to keep track of what compartments are
empty. He refers to the diagram below when he writes

Ascertain what each given proposition denies, and then put some kind of mark
upon the corresponding partition in the figure. The most effective means of doing

this is just to shade it out. For instance, the proposition 'All X is Y' is interpreted to mean that there is no such class of things in existence as 'X that is not-Y' or $X\overline{Y}$. All, then, that we have to do is to scratch out that subdivision in the two-circle figure thus,

"

He goes on to do examples using three and four propositions, and to describe in considerable detail the process for using his diagrams. Finally, he speculates on the design of a "logical machine," based on an electric abacus that someone named Prof. Jevons built. He even gives schematic diagrams for the construction of such a machine for four propositions. It is not clear whether the machine was ever built.

Venn says a subdivision exists or does not exist, where we would probably say it is empty or non-empty. He also "scratches out" the ones that do not exist, where we usually fill in the ones that are non-empty. Still, his diagrams look considerably more like modern diagrams, 120 years after Venn, than they look like Euler's circles, 120 years before Venn. Venn may begin where Euler ends, but he does add important new ideas of his own.

Venn himself might say that if X is the mathematics Euler did, and if Y is the mathematics Venn did, then surely $X\overline{Y}$, and indeed XY, but also $\overline{X}Y$.

References

[D] Dunham, William, *Euler The Master of Us All*, MAA, Washington, D. C., 1999.

[E] Euler, Leonhard, *Letters of Euler on Different Subjects in Natural Philosophy*, Henry Hunter, tr., Harper, New York, 1833, reprint edition Arno Press, 1975 and Edinburgh, 1823.

[R] Ruskey, Frank, A Survey of Venn Diagrams, *Electronic Journal of Combinatorics*, February 2, 1997 [ed March 15, 2001].

[V] Venn, John, "On the Diagrammatic and Mechanical Representation of Propositions and Reasonings", *The London, Edinburgh, and Dublin Philosophical Magazine and Journal of Science* [Fifth Series] July 1880, 1–18.

16

Knight's Tour

(April 2006)

It is sometimes difficult to imagine that Euler had a social life, but it is not surprising that he could find mathematics in what other people did for fun. He begins the article we are considering this month by writing:

> I found myself one day in a company where, on the occasion of a game of chess, someone proposed this question:

> *To move with a knight through all the squares of a chess board, without ever moving two times to the same square, and beginning with a given square.*

This is the problem now known as the Knight's Tour, and is an early special case of a Hamiltonian path on a graph, a problem that still occupies graph theorists.

Euler wrote this in the early 1750s, a time when a chess fad swept the courts of Europe. In 1751 the great chess master and good composer François-André Danican Philidor (1726–1795), whose games are still studied today, played before Frederick the Great at Potsdam and went on to visit Berlin. Euler might have met Philidor, or maybe not. Either way, it seems that Euler caught the Chess Bug, too. There are stories that he took up the game but was disappointed with how well he played. So he took some lessons and became "very good." I have been unable to verify these stories, and none of his chess games seem to have survived, so it is hard to know how good he might have been.[1]

Euler apparently wrote this article in 1758, though he had mentioned the Knight's tour in a letter to Goldbach in 1757. [JW] The article was published in the 1759 volume of the Berlin *Mémoires*, which, because of the Seven Years War (1756–1763), was not actually published until 1766. It was published again in 1849 in a posthumous collection of Euler's works. He mentions that this paper is based on "a particular idea that Mr. Bertrand[2] of Geneva gave me." After this paper, Euler did not return to mathematical problems in chess. He came very close, though. Knight's tours are closely related to a

---

[1]There are also stories that Euler composed some music, or maybe designed an algorithm for composing music, and that the results were awful. I have been unable to find any substance to these stories, either.

[2]Louis Bertrand (1731–1812)

kind of magic square called "pandiagonal," and Euler wrote about pandiagonal magic squares in 1779, when he wrote *Recherches sur un nouvelle espèce de quarrés magiques* (Researches on a new kind of magic squares) [E530]. This is a very long paper, but I cannot find that Euler mentions its connections to the Knight's tour.

A bit later in E309, Euler writes:

3. To make this question a bit clearer, I show here a route where, in beginning in one corner of the chess board, one moves through all the squares:

42	59	44	9	40	21	46	7
61	10	41	58	45	8	39	20
12	43	60	55	22	57	6	47
53	62	11	30	25	28	19	38
32	13	54	27	56	23	48	5
63	52	31	24	29	26	37	18
14	33	2	51	16	35	4	49
1	64	15	34	3	50	17	36

Figure 1. First example of an open Knight's tour

The numbers here mark the order of the squares that the Knight visits. In this example, the Knight starts in the lower left-hand corner, and finishes in the square just to the right of the starting point. This method of describing a tour is a bit hard to follow, so we will substitute a more modern and more graphical notation for the same tour, though Euler did not use this kind of notation:

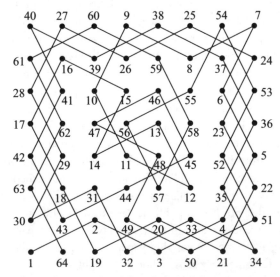

Figure 2. A graphical description of the same tour.

Since the Knight cannot move directly from its ending position back to its starting position, Euler says that this tour is "not re-entrant upon itself." We would call it an "open" tour or a Hamiltonian path, as opposed to a "closed" tour or a Hamiltonian circuit.

He gives us an example of a closed tour, "a route that is re-entrant upon itself," and notes that this gives a great many equivalent tours, starting this tour in any square, and traversing the numbers either forward or backward.

After these introductory comments and examples, the paper can be divided into several parts.

In paragraphs 9 to 14, he shows how new tours can be made from old ones by a technique that reconnects some of the steps in the path. This is probably the technique that he had learned from Bertrand.

Paragraphs 15 to 17 are devoted to using the reconnecting technique to extend a path that does not visit all the squares into a tour. Since the resulting tour is likely to be open, he spends paragraphs 18 to 24 showing how to reconnect an open tour to make it into a closed tour. This technique involves a lot of branching, and probably isn't very computationally efficient.

Starting with paragraph 25, Euler looks for tours with certain kinds of symmetries, like visiting first one half of the board, then the other half. Paragraphs 35 to 41 consider tours on boards that are not standard 8x8 chess boards. He looks at rectangular boards in paragraphs 42 and 43, and his last paragraph gives examples of four tours on boards that are shaped like crosses.

To get a sense of the main technique of the paper, rather than look at all the details, let's look at how Euler completes an incomplete tour.

Euler chooses the example in Figure 3, starting in the lower left corner and ending at step 62, missing the squares labeled a and b.

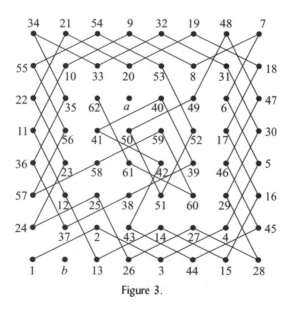

Figure 3.

First, make a list of the squares that can be reached from the last square, number 62:

9, 53, 59, 61, 23, 11, 55 and 21.

Now, also make a list of the squares that can be reached from the missing square a:

$$32, \; 8, \; 52, \; 42, \; 58, \; 56, \; 10 \text{ and } 54.$$

We notice that there are some squares in the first list for which the next square also appears on the second list, 9 and 10, 53 and 54, and 55 and 56. Euler could have selected any of these pairs, but he picks the pair 9 and 10 to reconnect the path.

Euler revises the path as follows. Instead of going from 9 to 10, he goes from 9 to 62. Then he traverses the path backwards until he gets to square 10, and from there he can reach the missing square a. He writes the new path as:

$$1 \; \ldots \; 9 - 62 \; \ldots \; 10 - a.$$

The resulting path is shown in Figure 4, with the connection from 9 to 10 that Euler removes shown with a dashed line, and the two new connections, 9 to 62 and 10 to a, shown with thick lines.

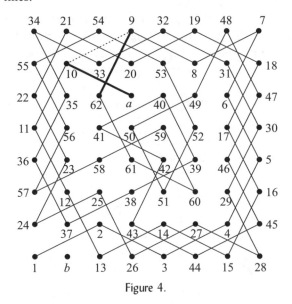

Figure 4.

This leaves b still to be added to the tour. Euler doesn't tell us why, but he doesn't try to connect b at the beginning of the tour. That would not be immediate. From b we can reach squares 57, 25 and 43, and 1 can reach only 2 and 12, and none of the resulting pairs are consecutive.

So, Euler lists again all the squares that can be reached from the new last square a:

$$32, \; 8, \; 52, \; 42, \; 58, \; 56, \; 10 \text{ and } 54.$$

The squares that can be reached from b are

$$57, \; 25 \text{ and } 43,$$

and again we have a pair of consecutive squares, 58 in the first list, which, in the revised path is followed by 57 in the second list. Note the complication, that we have to refer to the new path that Euler writes $1 \ldots 9 - 62 \ldots 10 - a$, and that in this path, 9 and 10 would not have been consecutive, but 9 and 62 are.

So, Euler makes another transformation. He disconnects square 58 from square 57, and instead connects it to square a. Then he traverses the path from a to 57 in the opposite direction, and connects 57 to the missing square b. He writes this new path as

$$1 \ldots 9 - 62 \ldots 58 - a - 10 \ldots 57 - b.$$

We show the new path in Figure 5, using dashes and thick lines as before.

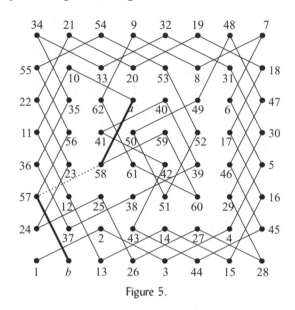

Figure 5.

This example hides some of the difficulties that can arise. For example, we usually see chess boards colored alternating red and black squares. A knight's move always takes it from a square of one color to a square of the other color. If we have a partial path that starts and ends on squares of the same color, then we cannot attempt to complete the path with a move to a square of that same color.

Further, if we have a partial path that starts and ends on squares of opposite colors, then we must try to add a missing square to the appropriately colored end of the path. Euler mentions these parity issues later in the paper when he is talking about tours on rectangles, but he does not mention it in this part of the paper.

Now that Euler has completed his partial tour to construct an open one, he wants to show us how to transform the open tour into a closed tour. His first step is to renumber the squares in their "natural order," that is the order in which this tour visits the squares instead of the order they were visited before he completed the tour. This relabeled tour is in Figure 6.

The process of closing an open tour is quite complicated, and this particular tour is more complicated than some because the starting point and the end point are near a corner, so there aren't as many ways to transform the tour. We will only summarize Euler's calculations.

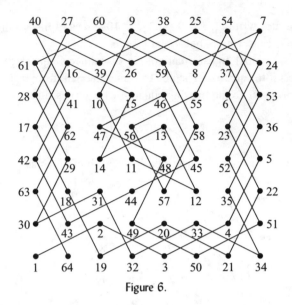

Figure 6.

Euler begins by listing the squares that can be reached from square 64:

63, 31 and 49.

Transposing at 64 does not result in a new tour, so he creates two new tours by transposing 64 first with 31 and then with 49. He names his new tours I and II, and describes them:

I. 1 ... 31 − 64 ... 32,

II. 1 ... 49 − 64 ... 50.

Now he reverses these tours, not changing their names but describing them as

I. 32 ... 64 − 31 ... 1,

II. 50 ... 64 − 49 ... 1.

The way Euler does it makes the calculations slightly easier because the two tours have the same endpoint. Now the last square, 1, connects to 2 and to 18. Transposing at 2 doesn't change anything, so he transposes at 18 and gets two new tours:

A. 32 ... 64 − 31 ... 18 − 1 ... 17,

B. 50 ... 64 − 49 ... 18 − 1 ... 17.

Both of these tours end at square 17, which, in turn, connects to squares

16, 10, 14 and 18.

He makes the four transpositions that change anything, and gets new tours C, D, E and F, starting at squares 32 or 50, and ending at 11 or 15. He makes all possible transformations at square 11 and gets tours G, H, I, K, L, M, N, O, P and Q, (note that he skips J because at the time I and J were the same letter) and also all transformations at square 15 to get tours g, h, i, k, l, m, n, o, p, q, r and s. Finally, he finds that one of the transformations that arise from tour G leads to a closed tour, and he finds the closed tour:

$$32 \ldots 42 - 47 \ldots 64 - 31 \ldots 18 - 1 \ldots 10 - 17 \ldots 11 - 46 \ldots 43.$$

He rewrites this so that it begins with square 1, and gets the final form of his closed tour:

$$1 \ldots 10 - 17 \ldots 11 - 46 \ldots 43 - 32 \ldots 42 - 47 \ldots 64 - 31 \ldots 18.$$

This method of closing an open path certainly is not computationally efficient, but that may be because it is a difficult problem, and there may be no better algorithm.

George Jelliss has written a nice history and a plethora of results on the Knight's tour and related topics. Readers who want to know more should surely visit his site [J].

References

[E309] Euler, Leonhard, Solution d'une question curieuse que ne paroit soumise à aucune analyse, *Mémoires de l'académie des sciences de Berlin*, 15 (1759) 1766, pp. 310–337, also in *Commentationes arithmeticae* 1, 1849, pp. 337–355, reprinted in *Opera Omnia* Series I vol. 7, pp. 26–56. Available online at EulerArchive.org and at the Digital Library at Berlin-Brandenburgische Akademie der Wissenschaften, bibliothek.bbaw.de/ bibliothek-digital/digitalequellen/schriften/.

[E530] ——, Recherches sur un nouvelle espèce de quarrés magiques, *Verhandelingen uitgegeven door het zeeuwsch Genootschap der Wetenschappen te Vlissingen*, 9, Middelburg, 1782, pp. 85–239, also in *Commentationes arithmeticae* 1, 1849, pp. 302–361, reprinted in *Opera Omnia* Series I vol. 7, pp. 291–392. Available online at EulerArchive.org.

[J] Jelliss, George, "Knight's Tour Notes," www.ktn.freeuk.com/index.htm, March 10, 2006.

[JW] Juskevic, A. P., and E. Winter, eds., *Leonhard Euler und Christian Goldbach: Briefwechsel 1729–1764*, Akademie-Verlag, Berlin, 1965.

17

Derangements

(September 2004)

Euler worked for a king, Frederick the Great of Prussia. When the King asks you to do something, he's not really "asking." In the late 1740s and early 1750s, the King "asked" Euler to work on a number of practical problems. For example, the King had a party palace named Sans Souci. Euler was asked to design the hydraulics to run the fountains at the palace. He also asked Euler to do the engineering on a canal. Another time, when the King was running out of money, he asked Euler to calculate probabilities so the King could try to pay his debts by running a lottery.

At about the same time, Euler was turning his talents to analyzing ordinary and frivolous things. He solved the Königsburg Bridge Problem, and the Knight's Tour problem, as well as analyzing some lotteries other than the one the King asked about. Among these other problems was a card game, called "le jeu de rencontre," or "the game of coincidence." He reported his results in a paper, E201, published in the *Mémoires* of the Berlin Academy under the title "Calcul de la probabilité dans le jeu de rencontre." Richard Pulskamp's translation of this article is available on line [P], and the original, besides appearing in Series I Volume 7 of the *Opera Omnia*, is on line through the Euler Archive [EA] and the Berlin Academy [B].

Rencontre takes two players, whom Euler names A and B. Their descendents still populate mathematics problems worldwide. The players have identical decks of cards. They both turn over cards, one at a time and at the same time. If they turn over the same

CALCUL DE LA PROBABILITÉ
DANS LE JEU DE RENCONTRE,
PAR M. EULER.

card at the same time, there is a coincidence, and A wins. If they go all the way through the deck without a coincidence, then B wins.

The problem is to calculate the probability that A will win. The probability will, of course, depend on the number of cards in the decks. Euler takes this number to be m. The problem still appears in many modern texts on probability, and the solutions given usually resemble Euler's. Since Jakob Bernoulli's *Ars Conjectandi* had appeared in 1713, Euler has at his command many of the standard tools of probability. In particular, we assume that all of the $m!$ possible permutations of the deck of cards are equiprobable (Euler uses neither the notation $m!$, nor the term "equiprobable") and that the probability that A wins is the number of successful arrangements divided by the number of possible arrangements.

First, we get a couple of simplifying assumptions that do not cost us any generality. First, we assume that the cards have numbers, $1, 2, 3, \ldots, m$, rather than designs. Second, we assume that A turns over cards in the order $1, 2, 3, \ldots, m$, so that the outcome of the game depends only on the order of B's cards.

Euler proceeds in his classical expository style. He starts with the easiest examples, $m = 1, 2, 3$ and 4. He does the case $m = 4$ two different ways, with the second method providing the idea that leads to a general solution.

The case $m = 1$ is trivial; A wins. If $m = 2$, there are two arrangements for B's cards, $1, 2$ and $2, 1$. A wins in the first case and loses in the second, so the probability that A wins is $\frac{1}{2}$.

We get some hints that Euler has some interesting ideas when he shows us the case $m = 3$. He gives us the table below, enumerating the six possible orders for B's cards, and with some entries crossed out. Before we explain what Euler says about this table, the reader should try to figure out what the table is about on his/her own:

A	**1**	**2**	**3**	**4**	**5**	**6**
1	1	1	2	2	3	3
2	2	3	1	3	2	1
3	3	2	3	1	1	2

One thing to do would be to cross out all the 1's in row 1, all the 2's in row 2 and the 3's in row 3. Then, any column that still has all its numbers represents an arrangement of B's cards that results in a win for B. This would show that B wins twice, columns 4 and 6, and A wins four times.

That's not what the table does. Columns 1 and 2 describe games in which A wins on the first move, so Euler has crossed out all the outcomes after row 1; they don't matter. Column 5 represents the game in which A wins on the second move, so Euler has crossed out the outcome after row 2. Column 3 describes the game in which A wins on the last move. There's nothing left to cross out, so Euler explains in the text that Column 3 represents the game in which A wins on the last move.

This is our first clue to what Euler intends to count. He will calculate the number of ways that A can win on move i if there are m cards. So far, his table would look kind of like this:

		Number of cards		
		I	II	III
Number of ways that A can win on move number __	I	1	1	2
	II		0	1
	III			1

That is skipping ahead a bit, though. Euler sticks to form and considers the case $m = 4$. He gives the following table:

A	B																							
	1	2	3	4	5	6	7	8	9	10	11	12	13	14	15	16	17	18	19	20	21	22	23	24
1	1	1	1	1	1	1	2	2	2	2	2	2	3	3	3	3	3	3	4	4	4	4	4	4
2	2	2	3	3	4	4	3	3	4	4	1	1	4	4	1	1	2	2	1	1	2	2	3	3
3	3	4	4	2	2	3	4	1	1	3	4	3	1	2	2	4	1	4	2	3	3	1	1	2
4	4	3	2	4	3	2	1	4	3	1	3	2	3	4	2	3	4	2	4	1	3	2	1	1

It takes some squinting, but we see that there are 6 ways that A wins on the first move, 4 ways to win on the second, 3 ways to win on the third, and 2 ways to win on the last move. This would add another column of data to the table we made a little earlier.

Now, Euler sets out to figure out how the table works. He studies the case $m = 4$ and asks about the games in which A wins on the third card. He extracts from the table above all the games for which there is a 3 in row 3, and gets the following sub-table:

A	B						
1	1	1	2	2	4	4
2	2	4	4	1	1	2
4	4	2	1	4	2	1

We notice that this table is almost exactly like the first table, the table of outcomes for the 3-card game, but all the 3's have been changed to 4's (though, for no apparent reason, the columns have been rearranged a little bit.) From these, he takes away those games in which A wins on the first card (two games) or the second card (one game), and the three games that remain must be the ones in which A wins the 4-card game on the third card. Euler has discovered the seeds of a recurrence relation, by which the number of ways to win a 4-card game on a particular move depends on the number of ways of winning various 3-card games. It will take some notation to untangle it. Unfortunately, subscripts had not yet been invented, so Euler has to make do without them.

Suppose there are m cards in the deck, and that the total number of possible games is M. We know that $M = m!$, but the factorial notation hadn't been invented yet, either. Now, let

a be the number of cases for which A wins on the first move,

b be the number for which A wins on the second move,

c be the number for the third move,

etc.

Easy analysis shows that $a = \frac{M}{m}$.

Now, consider the game with $m + 1$ cards. Euler denotes by M', a', b', c', etc. the corresponding numbers for the larger game, and asks how the numbers for the $(m+1)$-card game are related to those for the m-card game.

Some parts are easy; $M' = M(m + 1)$ and $a' = \frac{M'}{m+1} = M$.

Now, there are M cases in which A turns over a 2 on the second card, but some of these must be excluded, since they are cases in which A has already turned over a 1 on the first card. The analysis Euler did on the reduced table tells us to look at the m-card games to find that there are a such arrangements, so that $b' = M - a$.

Likewise, there are M cases in which A turns over a 3 on the third card, but from these we must subtract those cases in which A has already won on the first or the second card. That is $c' = M - a - b$.

The pattern continues. We can write these relations in a simpler form:

$$a' = M$$
$$b' = a' - a$$
$$c' = b' - b$$

etc.

Euler uses these results to calculate the following table for up to 10 cards. This is the same table we derived ourselves for up to 3 cards.

	I	II	III	IV	V	VI	VII	VIII	IX	X
						NOMBRE DES CARTES				
a	1	1	2	6	24	120	720	5040	40320	362880
b	-	0	1	4	18	96	600	4320	35280	322560
c	·	·	1	3	14	78	504	3720	30960	287280
d	-	-	·	2	11	64	426	3216	27240	256320
e	·	-	·	·	9	53	362	2790	24024	229080
f	·	·	-	·	·	44	309	2428	21234	205056
g	-	-	·	·	·	·	265	2119	18806	183822
h	·	·	·	·	·	·	·	1854	16687	165016
i	·	·	·	·	·	·	-	· ·	14833	148329
k	-	·	-	·	·	·	-	· ·	· -	133496

The hard work is over, but Euler promised to calculate the probabilities, too. Let A be the probability that A wins on the first move of an $(n-1)$-card game, B the probability he wins on the second, C the probability he wins on the third, and so forth, and let N be the number of possible $(n-1)$-card games. That is, $N = (n-1)!$. Similarly, let A', B', C' and N' be the corresponding probabilities for an n-card game.

It is easy to see that $A = \frac{a}{N} = \frac{1}{n-1}$ and $A' = \frac{a'}{N'} = \frac{1}{n}$. Now, $B' = \frac{b'}{N'}$. But, $b' = a' - a$, and $N' = nN$, so

$$B' = \frac{b'}{N'} = \frac{a'-a}{N'} = \frac{a'}{N'} - \frac{a}{N'} = \frac{1}{n} - \frac{a}{nN} = \frac{1}{n} - \frac{1}{n(n-1)} = \frac{n-2}{n(n-1)}$$

Changing n's to $(n-1)$'s gives that

$$B = \frac{n-3}{(n-2)(n-1)}$$

Similar calculations show a clear pattern:

$$A' = \frac{1}{n}$$

$$B' = \frac{1}{n} - \frac{1}{n(n-1)}$$

$$C' = \frac{1}{n} - \frac{2}{n(n-1)} + \frac{1}{n(n-1)(n-2)}$$

$$D' = \frac{1}{n} - \frac{3}{n(n-1)} + \frac{3}{n(n-1)(n-2)} - \frac{1}{n(n-1)(n-2)(n-3)}$$

Numerators are rows from Pascal's triangle. Denominators are permutation numbers.

Euler sums the columns. The probability we seek, that player A wins on *some* move, is the sum of the n probabilities on the left-hand side, $A' + B' + C' + \cdots$. He sums the right-hand sides as columns, since the denominators match. This is easier than it looks, since he knows lots of identities about Pascal's triangle.

For the first column,

$$1 + 1 + \cdots + 1 = n,$$

so the first term of the sum will be $n/n = 1$.

For the second column,

$$1 + 2 + 3 + \cdots + n = \frac{n(n-1)}{2}$$

so the second term of the sum will be

$$\frac{-\left(\frac{n(n-1)}{2}\right)}{n(n-1)} = \frac{-1}{1 \cdot 2}$$

For the third column,

$$1 + 3 + 6 + \cdots + \frac{(n-1)(n-2)}{2} = \frac{n(n-1)(n-2)}{1 \cdot 2 \cdot 3}$$

so the third term of the sum will be $\frac{1}{1 \cdot 2 \cdot 3}$.

Wow! Numerators alternate 1 and -1, while denominators are factorials! So, the probability that A wins playing with an n-card deck is

$$1 - \frac{1}{2!} + \frac{1}{3!} - \frac{1}{4!} + \cdots \pm \frac{1}{n!}$$

As n grows, this converges rapidly to $1 - 1/e$. For $n = 10$, it is already accurate to six decimal places. It is an astonishing result.

Now, about our title, "derangements." In discrete mathematics, combinatorics and abstract algebra courses, we learn about permutations, one-to-one and onto functions from a set to itself. They have all sorts of wonderful properties; they form a group and they are fun to count.

A *derangement* is a special kind of permutation, σ, with no fixed points. That is, it never happens that $\sigma(x) = x$. As a permutation, *everything* gets moved. Derangements correspond to those rearrangements of the deck for which A wins the game of rencontre.

Derangements sometimes appear as "the hat-check problem." One (obviously rather dated) version goes like this:

> Ten men go into a restaurant and check their hats. As they are leaving, the lights go out, and each man gets a hat at random. What is the probability that at least one man gets his own hat?

This is obviously 10-card rencontre in disguise. Now that we know how Euler did it, we know that the answer, to at least six decimal places, is $1 - 1/e$.

References

[B]   Digitalisierte Akademieschriften und Schriften zur Geschichte der Königlich Preussischen Akademie der Wissenschaften (Digital library of the Royal Prussian Academy of Sciences, 1700–1900) bbaw.de/bibliothek/digital/.

[EA]   The Euler Archive, EulerArchive.org.

[E201] Euler, Leonhard, "Calcul de la probabilité dans le jeu de rencontre", *Mémoires de l'Académie des Sciences de Berlin*, 7 (1751) 1753, pp. 255–270, reprinted in *Opera Omnia* Series I vol. 7 pp. 11–25. Available online at EulerArchive.org.

[P]   Pulskamp, Richard, "Leonhard Euler on Probability and Statistics," cerebro.xu.edu/math/Sources/Euler.html.

18

Orthogonal Matrices

(August 2006)

Jeff Miller's excellent site [M] "Earliest Known Uses of Some of the Words of Mathematics" reports:

> The term **MATRIX** was coined in 1850 by James Joseph Sylvester (1814–1897):
>
> > For this purpose we must commence, not with a square, but with an oblong arrangement of terms consisting, suppose, of m lines and n columns. This will not in itself represent a determinant, but is, as it were, a Matrix out of which we may form various systems of determinants by fixing upon a number p, and selecting at will p lines and p columns, the squares corresponding of pth order.
>
> The citation above is from "Additions to the Articles 'On a new class of theorems', and 'On Pascal's theorem'," *Philosophical Magazine*, pp. 363–370, 1850. Reprinted in Sylvester's *Collected Mathematical Papers*, vol. 1, pp. 145–151, Cambridge (At the University Press), 1904, page 150.

On the subject of orthogonal matrices, he writes

> The term **ORTHOGONAL MATRIX** was used in 1854 by Charles Hermite (1822–1901) in the *Cambridge and Dublin Mathematical Journal*, although it was not until 1878 that the formal definition of an orthogonal matrix was published by Frobenius (Kline, page 809).

Imagine my surprise when I was browsing Series I Volume 6 of Euler's *Opera Omnia* trying to answer a question for Rob Bradley, President of The Euler Society. This is the volume *Ad theoriam aequationum pertinentes*, "pertaining to the theory of equations." In the middle of that volume I found an article [E407] with the unremarkable title "Algebraic problems that are memorable because of their special properties," *Problema algebraicum ob affectiones prorsus singulares memorabile*. Usually, it seems, when Euler calls a problem "memorable," I don't agree. This was an exception.

Euler asks us to find nine numbers,

$$A, \quad B, \quad C,$$
$$D, \quad E, \quad F,$$
$$G, \quad H, \quad I$$

that satisfy twelve conditions:

1.	$A^2 + D^2 + G^2 = 1$	7.	$A^2 + B^2 + C^2 = 1$
2.	$B^2 + E^2 + H^2 = 1$	8.	$D^2 + E^2 + F^2 = 1$
3.	$C^2 + F^2 + I^2 = 1$	9.	$G^2 + H^2 + I^2 = 1$
4.	$AB + DE + GH = 0$	10.	$AD + BE + CF = 0$
5.	$AC + DF + GI = 0$	11.	$AG + BH + CI = 0$
6.	$BC + EF + HI = 0$	12.	$DG + EH + FI = 0$

If we regard the nine numbers as a 3×3 matrix

$$M = \begin{bmatrix} A & B & C \\ D & E & F \\ G & H & I \end{bmatrix}$$

then conditions 1, 2, 3, 10, 11 and 12 are exactly the conditions that make $MM^T = I$. In modern terms, this makes M an orthogonal matrix.

Note that the other six conditions make $M^T M = I$, another characterization of orthogonal matrices.

Of course, Euler doesn't *know* these are orthogonal matrices. When he wrote this paper in 1770, people didn't use matrices to do their linear algebra. As Jeff Miller's site suggests, Euler was doing this 80 years before these objects joined the mathematical consciousness.

What, then, was Euler thinking when he formulated this problem? We can only speculate. It was probably something more than that he admired the pretty patterns, but something less than that he sensed the profound power of operations on such arrays of numbers.

Let's look at Euler's paper. First he notes that we have 12 equations, but only nine unknowns, so there is a chance that the problem has no solution. Rather than showing that the system of equations has a solution by giving an example of a solution (say $A = E = I = 1$, all the rest of the unknowns equaling zero), he offers a theorem:

Theorem. *If nine numbers satisfy the first six conditions given above, then they also satisfy the other six.*

Euler himself describes the proof of this theorem as *calculos vehementer intricatos*, "vehemently intricate calculations," and we won't describe them in any significant detail. He does do an interesting step at the beginning, though.

Euler re-writes conditions 4, 5 and 6 as

4.	$AB = -DE - GH$
5.	$AC = -DF - GI$
6.	$BC = -EF - HI$

Then he asks us to multiply equation 4 by equation 5, then divide by equation 6 to get, as he writes it,

$$\frac{4 \cdot 5}{6} : \frac{AABC}{BC} = AA = -\frac{(DE + GH)(DF + GI)}{EF + HI}.$$

The notation $\frac{4 \cdot 5}{6}$ is not an arithmetic operation, but an *ad hoc* notation for an algebraic operation on equations 4, 5 and 6. Euler does this one or two times in other papers, but this is the only time he uses such a notation in this paper.

After three pages of such calculations, Euler eventually derives all of the conditions 7 to 12 from conditions 1 to 6, thus proving his theorem.

Now Euler turns to the "solution of the problem that was proposed at the beginning." Condition 1 (or 7) guarantees that A is between -1 and 1, so it has to be the cosine of something. Let $A = \cos . \zeta$. (Note that Euler still uses cos. as an abbreviation for "cosine," hence the period. Now that we've mentioned it, we'll write it the modern way. We'll also write $\sin^2 \zeta$ where Euler wrote $\sin . \zeta^2$.)

Then, from conditions 1 and 7, we get

$$DD + GG = 1 - AA = 1 - \cos^2 \zeta = \sin^2 \zeta$$

and similarly

$$BB + CC = \sin^2 \zeta.$$

These equations will be satisfied by taking

$$B = \sin \zeta \cos \eta, \quad C = \sin \zeta \sin \eta, \quad D = \sin \zeta \cos \theta, \quad \text{and} \quad G = \sin \zeta \sin \theta.$$

Let's check the score. We have six equations, nine unknowns, and we've made three arbitrary decisions by choosing ζ, η and θ. Euler knows enough linear algebra (see, for example, [S 2004]) to suspect that this will determine a unique solution. Indeed, they determine A, B, C, D and G. With half a page of calculations, he first finds that

$$E = \sin \eta \sin \theta - \cos \zeta \cos \eta \cos \theta$$

and

$$I = \cos \eta \cos \theta - \cos \zeta \sin \eta \sin \theta$$

and then that

$$F = -\cos \eta \sin \theta - \cos \zeta \sin \eta \cos \theta$$

and

$$H = -\sin \eta \cos \theta - \cos \zeta \cos \eta \sin \theta.$$

This solves the problem. In typical Eulerian fashion, though, he doesn't stop there. Instead, he develops some slightly more efficient techniques and goes on to solve analogous problems for 4×4 and 5×5.

Almost as if Euler did not want us to believe that he was actually doing modern linear algebra, Euler's last problem is to find a 4×4 array of integers

$$
\begin{array}{cccc}
A & B & C & D \\
E & F & G & H \\
I & K & L & M \\
N & O & P & Q
\end{array}
$$

(he skipped "J" on purpose) satisfying the 12 orthogonality equations

$$AE + BF + CG + DH = 0$$

$$AI + BK + CL + DM = 0$$

etc.

and the additional condition on the sums of the squares of the numbers on the two diagonals,

$$A^2 + F^2 + L^2 + Q^2 = D^2 + G^2 + K^2 + N^2.$$

He demonstrates how to find two different solutions. One is

68	−29	41	−37
−17	31	79	32
59	28	−23	61
−11	−77	8	49

where the dot product of any row (column) with any other row (column) is zero, and the sum of the squares of 68, 31, −23 and 49 equals the sum of the squares of −37, 79, 28, −11. They both equal 8415.

His other solution is

73	−85	65	−11
−53	31	107	41
−89	−67	1	−67
−29	−65	−35	103

where the sums of the squares on the diagonals are 16,900. Euler's methods are typical of his work in Diophantine equations, and would allow us to generate as many solutions as we want.

References

[E407] Euler, Leonhard, Problema algebraicum ob affectiones prorsus singulares memorabile, *Novi Commentarii academiae scientiarum Petropolitanae* 15, (1770) 1771, pp. 75–101, reprinted in *Opera Omnia* Series 1, vol. 6, pp. 287–315. Available online at EulerArchive.org.

[M]   Miller, Jeff, Earliest Uses of Symbols in Number Theory, members.aol.com/jeff570/nth.html, consulted June 13, 2006.

[S2004] Sandifer, Ed, Cramer's Paradox, *How Euler Did It*, MAA Online, August 2004. Reprinted in this volume, pp. 37–42.

Part IV

Analysis

19

Piecewise Functions
(January 2007)

This year, 2007, marks the 300th anniversary of Euler's birth on April 15, 1707. We begin our celebration of Euler's birthday by discussing one of Euler's most fundamental contributions to mathematics, the idea of a function.

The word "function" comes to us from the Latin *functio*, meaning a performance, an event or an activity, not, as we might hope, from the German *der Funke*, a spark or a glimmer. (The colloquial "funky" comes from the German.)

Today, functions are one of the central objects in mathematics. David Hilbert told us, "Besides the concept of number, the concept of function is the most important one in mathematics." [T] On the other hand, Hilbert's student, Hermann Weyl, wrote, "Nobody can explain the function concept." [T]

The Ancients knew some of the relations between curves and algebraic expressions. Both Apollonius and Archimedes, for example, knew how the shape of a parabola was related to the algebraic expression $ay = x^2$, though they didn't use algebraic notation to express the relation. They were philosophically and notationally unable to make sense of an expression like $y = x^2$ because one object in the expression, y, is a length and the other, x^2 represents an area. They regarded relations like $ay = x^2$ as properties of curves, and not as definitions of the curves themselves, and they called such properties *symptoms*.

In the early 1600s, Descartes devoted a big piece of his *Geometria* to giving meaning to nonhomogeneous expressions like $y = x^2$. Soon the idea developed that every curve had an associated algebraic expression of some sort, but the formula was still regarded as a property of the curve. Formulas were not yet stand-alone objects.

Functions gradually earned their own identity as the 18th century progressed. In 1696, when L'Hôpital wrote *Analyse des infiniment petits pour l'intelligence des lignes courbes*, the world's first calculus book, he wrote about curves, and a curve existed if it could be constructed by some mechanical or geometric process. Fifty years later, Euler wrote the *Introductio in analysin infinitorum* from the point of view of functions, and a curve existed if it could be described by an analytic expression. In fact, Rob Bradley [B] describes an interesting story contrasting the two ideas of what makes a curve. L'Hôpital had described

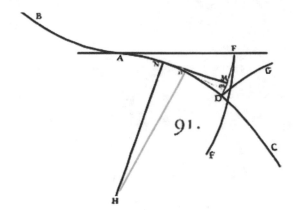

a "cusp of the second kind" sometimes called a "bird's beak." L'Hôpital was studying involutes of curves like the curve $BANDC$ shown in his Figure 91 above. The curve has an inflection point at A. The involute of the curve is shown as the awkwardly named curve $DMFF$, which, at its point F corresponding to the inflection point A, has a cusp for which both branches curve the same direction, like a bird's beak. Hence the name.

In 1696, people had no problem accepting that such curves existed. There was a clear mechanical construction. By 1740, though, people weren't so sure, since they couldn't seem to find an analytical representation of such curves. In 1748 in the *Introductio*, Euler gave a formula, and the bird's beak was restored. It was curious that people believed their formulas more than they believed their eyes.

Euler was a bit like Hermann Weyl when it came to the function concept itself. Euler knew what he wanted functions to *do*, but he sometimes struggled to articulate what they *are*. Early on, a function was an analytic expression describing a curve. In an expression like $x^2 + y^2 = 1$, x is a function of y, but y is also a function of x, since knowing one, x or y, we can determine the other. Euler also allowed multi-valued functions. For example, in the expression $y = x^2$, y is a single-valued function of x, but x is a multi-valued function of y.

Euler also accepted solutions to differential equations as functions, even if those solutions might not be written down explicitly. Of course, he was quite unaware of the exotic "pathological" functions that Weierstrass and Dirichlet would describe in the 19th century.

Euler was not always consistent as he struggled to refine the concept of a function. As an example of this struggle, we will take a closer look at Euler's thoughts about what we now call "piecewise defined functions."

Euler usually thought that a function had to be defined by the same analytical expression everywhere. Since he did not have any notation for the absolute value function, perhaps the best-known piecewise-defined function, he never had cause to realize that a function as natural as the absolute value function is actually defined piecewise. He occasionally came across the absolute value function disguised as $\sqrt{x^2}$, but when he did, he was always interested in other issues.

He began rejecting piecewise functions early in his career. Two of his earliest papers, E3, *Methodus inveniendi traiectorias reciprocas algebraicas* and E5, *Problematis traiectoriarum reciprocarum solutio*, deal with the now forgotten and misunderstood topic of "reciprocal trajectories," curves with a peculiar kind of symmetry that people sometimes

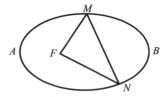

Figure 1.

incorrectly believe has something to do with ballistic trajectories. Reciprocal trajectories are somewhat esoteric, and rather than investing the time to explain them, we'll jump forward a few years to the fruits of one of Euler's shortest papers.

In 1745, Euler sent a short note to *Nova Acta eruditorum* forwarding a problem posed anonymously by Christian Goldbach. The note became an eight-line "paper" [E79] titled "A problem of geometry proposed publicly by an anonymous geometer," probably Euler's shortest paper and maybe one of the shortest mathematics papers anyone ever wrote. In E79, Euler and Goldbach, referring to Figure 1, ask what curves like $AMBN$ there might be with the property that there is a point F from which any ray, like FM, reflected twice, returns to the point F.

Ellipses have this property. The point F can be either of the foci of the ellipse. It is a familiar property of ellipses that any ray from one focus reflects to pass through the other focus. There it becomes a ray from a focus, so it will reflect again and return to the first focus. Goldbach and Euler ask if there are any other such curves, or if this is in itself a defining property of an ellipse?

In E79, Euler only posed the problem, but he solved it two years later in E106, "Solution to the catoptric problem in *Novis Actis Eruditorum Lipsiensibus* proposed in November 1745." He found that there were, indeed, curves other than the ellipse with this special property, and then, in typical Eulerian style, he turned to variations of the same problem. He looked at a problem that is projectively related to the ellipse-like problem he started with. He sought to find if the parabola is the only curve like $FMBmf$, shown in Figure 2, with the property that rays like CM, parallel to the axis of the curve AB reflected twice, as Mm, then mc, will give a ray mc parallel to the original.

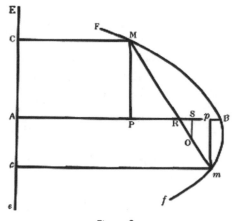

Figure 2.

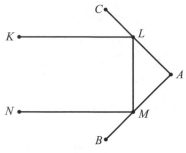

Figure 3.

Much like ellipses, parabolas have this property. The reflection of the ray CM will pass through the focus of the parabola, shown in the figure as the point R on the axis AB of the parabola. Euler asks if there are any other such curves, and, of course he finds some. In the course of his analysis, though, he explicitly rejects curves like the one shown in Figure 3, where the two branches are not described by the same analytic expression. In this figure, the branch AC is perpendicular to the branch AB, and the rays KL and MN are parallel to the bisector of angle CAB.

There was a philosophical basis for rejecting curves like the absolute value function. Leibniz championed something usually called the "principle of continuity," but since the word "continuity" means something different to mathematicians, we'll call it the "principle of continuation."[1] Broadly speaking, the principle of continuation says that similar things will behave similarly. Leibniz summarized the principle, writing "Nature makes no leaps." We have seen Euler use the principle of continuation before when he does arithmetic with infinite and infinitesimal "numbers" just as if they were ordinary finite numbers.

In the present problems, the principle of continuation tells us that the solution to a "natural" problem will not have any "leaps" in it, and it will be described by a single analytic expression.

Given this world view, it would be surprising if Euler proposed a piecewise function as the solution to a "natural" problem. Yet he did exactly that when he analyzed ballistic trajectories. In the article we described in last month's column[2] [E217] he tells us that the forces acting on a cannonball (*not* neglecting air resistance) give different differential equations for the ascending branch than for the descending branch. In particular, Euler takes x and y coordinates as usual, s to be arc length, t to be time, and c is a parameter describing the properties of air. The variable v, though, might be confusing to the modern reader. It is the height from which an object would have to be dropped to have the same speed as the cannon ball has at time t. Hence, v is a length, not a velocity or a speed, and because of his choice of units, the speed, $\frac{ds}{dt}$ is given by

$$\frac{ds}{dt} = \sqrt{v}.$$

Since air resistance is taken to be proportional to the square of the speed, this makes air resistance proportional to v itself. Euler also takes α to be the acceleration due to gravity.

---

[1]Of course, the name "principle of continuation" has its own meaning in other contexts. For example, to geologists, it describes a property of layers of sediments.

[2]"Cannonball curves," December 2006.

With this notation in place, Euler resolves forces and finds that the acceleration in the x direction is given by the same differential equation,

$$\frac{2d\,dx}{t\,d^2} = -\frac{v\,dx}{c\,ds}$$

whether the cannonball is ascending or descending. In what seems to be a violation of the law of continuation, the acceleration is given by

$$\frac{2d\,dy}{dt^2} = \alpha - \frac{v\,dy}{c\,ds}$$

when the cannonball is ascending, but it is given by

$$\frac{2d\,dy}{dt^2} = \alpha + \frac{v\,dy}{c\,ds}$$

when it is descending.

I suspect that Euler was not thinking about the law of continuation when he wrote this. If he had, though, he might have tried to explain it by noting that at the apex, where the trajectory changes from its ascending branch to its descending branch, the factor $\frac{dy}{ds}$ gradually vanishes and reappears. The leap isn't in nature, but in our notation.

We can't let Euler off the hook that easily, though. I was careful above to describe the symbol c as a parameter, not a constant. Euler takes e^3 to be the volume of water with the same mass as the cannonball and d to be the diameter of the cannonball. Then he tells us that if the speed of the projectile is not too fast, then

$$c = \frac{2133e^3}{d\,d}.$$

However, he seems to have done some experiments and concluded that "if the movement is so rapid that air cannot immediately occupy the space behind the globe, then the globe will leave behind itself a kind of void space, and so for that instant the globe will be subject to the full pressure of the atmosphere, which will not be counterbalanced by an equal pressure from behind, and so it is clear that the resistance will be augmented by the entire pressure of the atmosphere on the part at the front of the globe." Euler calculates that this changes the air resistance from

$$\frac{v}{c} \qquad \text{to} \qquad \frac{v}{c} + \frac{6666k}{4c},$$

where k is the air pressure measured in feet of water.

This is a sudden change in force that Euler's data indicates occurs when $v > 28050$ feet. Euler tells us this translates to 1325 feet per second. Modern theory puts this change close to the speed of sound, 1087 feet per second. Euler knew the speed of sound fairly accurately, so he apparently didn't understand how the speed of sound is related to this phenomenon.

Euler apparently did not try to reconcile this sudden change in force with the principle of continuation. After all, the theory worked to describe trajectories. The void space behind the cannonball seemed to explain the phenomenon, even if the analytic representation makes that troublesome leap.

I think that, in not committing too strongly to the principle of continuation, Euler displayed an admirable lack of rigor that left the concept of a function with enough flexibility that it could evolve into the foundation of mathematics it has become today.

References

[B]   Bradley, Robert E., The Curious Case of the Bird's Beak, *International Journal of Mathematics and Computer Science*, vol. 1, no. 2, (2006), pp. 243–268.

[E79]   Euler, Leonhard, Problema geometricum propositum publice ab anonymo geometra, *Nova acta eruditorum*, 1743 pp. 523, reprinted in *Opera Omnia* Series I vol. 27 p. 50. Available online at `EulerArchive.org`.

[E106]   ——, Solutio problematis catoptrici in *Novis Actis Eruditorum Lipsiensibus* pro mense Novembri A. 1745 propositi, *Nova acta eruditorum*, 1748, pp. 27–46, 61–75, 169–184, reprinted in *Opera Omnia* Series I vol. 27 pp. 78–129. Available online at `EulerArchive.org`.

[E217]   ——, Recherches sur la véritable courbe qui décrivent les corps jettés dans l'air ou dans un autre fluide quelconque, *Mémoires de l'académie des sciences de Berlin*, 9 (1753) 1755, pp. 321–362. Reprinted in *Opera Omnia* Series II vol. 14 pp. 413–447. Available online at `EulerArchive.org`.

[T]   Thiele, Rüdiger, The Function Concept, from an address accepting the Lester R. Ford award, MathFest, Providence, RI, August 2004.

20

Finding Logarithms by Hand

(July 2005)

Today, it is just as easy to take a square root as it is to find a logarithm. You just find your calculator, turn it on, and press a few buttons. The first step of that process usually takes the longest. In 1748, though, when Euler published the *Introductio in analysin infinitorum*, most mathematicians and scientists were quite good at taking square roots by hand, but logarithms required difficult analysis or a book of tables. The seven-place tables of Briggs and Vlacq had been available for over a hundred years, but the ten place tables of Jurij Vega were still fifty years in the future.

In the *Introductio*, perhaps Euler's most interesting and influential work, Euler did not append tables of anything, be it logarithms, square roots, or trigonometric functions. So, when he covered this material (except for square roots, which he assumed we already know) he wanted to make sure his readers knew how to find the values of the functions, even if in practice they would be looking them up in tables. For the trigonometric tables, Euler gave fast converging series and a few tricks involving trig identities that gave values to 20 or more decimal places in just a dozen or so steps. In Chapter VI, "On Exponentials and Logarithms," though, he gives an algorithm that does not use series.

Euler's algorithm is based on the elementary properties of logarithms and the fact that $\log x$ is a monotone increasing function of x.

Notation in 1748 was a bit awkward by modern standards. Euler always wrote $l\,x$ where we would write $\log x$ or $\ln x$, and he always has to tell us what the base of the logarithm is. We will use the modern notation.

Euler begins his description of the algorithm by telling us "putting $\log y = z$ & $\log v = x$, then $\log \sqrt{vy} = \frac{x+z}{2}$." That is, geometric means of the arguments correspond to arithmetic means of the logarithms.

Euler prepares us for his algorithm by explaining, "if a given number b lies between limits a^2 and a^3, the logarithms of which are 2 and 3 respectively, then we will find the value of $a^{2\frac{1}{2}}$, that is $a^2 \sqrt{a}$, and then b must lie either between the limits a^2 and $a^{2\frac{1}{2}}$, or between $a^{2\frac{1}{2}}$ and a^3." Euler is setting us up for what we now call a binary search.

For his example, Euler takes $a = 10$, so we will be finding base ten logarithms. He

takes $b = 5$, and plans to find $\log 5$. He notes that 5 is between 1 and 10, so $\log 5$ is between 0 and 1.

Now, the geometric mean of 1 and 10 has as its logarithm the arithmetic mean of 0 and 1. With this, Euler begins a binary search. He writes

$$A = 1.000000 \qquad \log A = 0.000000$$
$$B = 10.000000 \qquad \log B = 1.000000 \qquad C = \sqrt{AB}$$
$$C = 3.162277$$

Now, using the log $\sqrt{vy} = \frac{x+z}{2}$ fact, since C is the geometric mean of A and B, its logarithm is the arithmetic mean of $\log A$ and $\log B$, that is 0.500000. He adds this to his table and gets

$$A = 1.000000 \quad \log A = 0.0000000$$
$$B = 10.000000 \quad \log B = 1.0000000 \quad C = \sqrt{AB}$$
$$C = 3.162277 \quad \log C = 0.5000000$$

Since 5 is between $B = 10.000000$ and $C = 3.162277$ (and not between A and C), $\log 5$ is between $\log B$ and $\log C$. He repeats his first step and takes D to be the geometric mean of B and C, so its log is the arithmetic mean. Euler's table grows to be

$$A = 1.000000 \quad \log A = 0.0000000$$
$$B = 10.000000 \quad \log B = 1.0000000 \quad C = \sqrt{AB}$$
$$C = 3.162277 \quad \log C = 0.5000000 \quad D = \sqrt{BC}$$
$$D = 5.623413 \quad \log D = 0.7500000$$

Euler continues this process of bisection and geometric means until he runs out of letters. Eventually, he gets $X = 4.999997$, $Y = 5.000003$ and $Z = 5.000000$, and $\log z = 0.6989700$. Euler's whole table is shown below:

$$76 \qquad DE \; QUANTITATIBUS$$

Lib. I.				
$A = 1,000000;$	$lA = 0,$	0000000		fit
$B = 10,000000;$	$lB = 1,$	$0000000;$	$C = \sqrt{AB}$	
$C = 3, 162277;$	$lC = 0,$	$5000000;$	$D = \sqrt{BC}$	
$D = 5, 623413;$	$lD = 0,$	$7500000;$	$E = \sqrt{CD}$	
$E = 4, 216964;$	$lE = 0,$	$6250000;$	$F = \sqrt{DE}$	
$F = 4, 869674;$	$lF = 0,$	$6875000;$	$G = \sqrt{DF}$	
$G = 5, 232991;$	$lG = 0,$	$7187500;$	$H = \sqrt{FG}$	
$H = 5, 048065;$	$lH = 0,$	$7031250;$	$I = \sqrt{FH}$	
$I = 4, 958069;$	$lI = 0,$	$6953125;$	$K = \sqrt{HI}$	
$K = 5, 002865;$	$lK = 0,$	$6992187;$	$L = \sqrt{IK}$	
$L = 4, 980416;$	$lL = 0,$	$6972656;$	$M = \sqrt{KL}$	
$M = 4, 991627;$	$lM = 0,$	$6982421;$	$N = \sqrt{KM}$	
$N = 4 997 42;$	$lN = 0,$	$6987304;$	$O = \sqrt{KN}$	
$O = 5, 000052;$	$lO = 0,$	$6989745;$	$P = \sqrt{NO}$	
$P = 4, 998647;$	$lP = 0,$	$6988525;$	$Q = \sqrt{OP}$	
$Q = 4, 999350;$	$lQ = 0,$	$6989135;$	$R = \sqrt{OQ}$	
$R = 4, 999701;$	$lR = 0,$	$6989440;$	$S = \sqrt{OR}$	
$S = 4, 999876;$	$lS = 0,$	$6989592;$	$T = \sqrt{OS}$	
$T = 4, 999963;$	$lT = 0,$	$6989668;$	$V = \sqrt{OT}$	
$V = 5, 000008;$	$lV = 0,$	$6989707;$	$W = \sqrt{TV}$	
$W = 4, 999984;$	$lW = 0,$	$6989687;$	$X = \sqrt{WV}$	
$X = 4, 999997;$	$lX = 0,$	$6989697;$	$Y = \sqrt{VX}$	
$Y = 5, 000003;$	$lY = 0,$	$6989702;$	$Z = \sqrt{XY}$	
$Z = 5, 000000;$	$lZ = 0,$	$6989700;$		

Note that the logarithms are given to seven places, just as in the tables by Briggs and Vlacq.

Euler then shows how log 2 is easily found as $1 - \log 5$ and notes that with these two values it is now easy to find the logs of 4, 8, 16, 32, 64, etc., as well as 20, 40, 80, 25, 50, etc.

This lesson on finding logarithms by hand ends with this. In the next chapter, Chapter VII titled "Exponentials and Logarithms Expressed through Series," Euler does use the series methods we see more often today. These are similar to the series methods that Briggs had used over a hundred years earlier. We will briefly describe Euler's methods.

Euler has switched to the natural logarithm, which he still denotes $l\,x$, and mentions in the text that the base is e. First, Euler finds the well-known series

$$\ln(1 + x) = x - \frac{x^2}{2} + \frac{x^3}{3} - \frac{x^4}{4} + \frac{x^5}{5} - \text{ etc.}$$

Then, $\log(1-x)$ has almost the same series, but it does not alternate. Using the elementary properties of logarithms, we get to subtract the two series. This eliminates all the terms of even degree and gives

$$\ln\left(\frac{1+x}{1-x}\right) = \frac{2x}{1} + \frac{2x^3}{3} + \frac{2x^5}{5} + \frac{2x^7}{7} + \text{ etc.}$$

For small values of x, this last series converges rapidly. If we pick clever values of x, we can start to construct tables of logarithms. For example, if $x = \frac{1}{5}$, we get $\ln\frac{6}{4}$. Likewise, taking $x = \frac{1}{7}$ and $x = \frac{1}{9}$ gives the logs of $\frac{4}{3}$ and $\frac{5}{4}$, respectively. Finally, we combine these to get $\ln 2 = \ln\frac{3}{2} + \ln\frac{4}{3}$, $\ln 3 = \ln\frac{3}{2} + \ln 2$ and $\ln 4 = 2\ln 2$. Similar tricks give logarithms for 5, 6, 8, 9 and 10, all from just three series.

Note that we missed the logarithm of 7. This requires a famous trick that Briggs used for base 10 logarithms. Now Euler repeats it for natural logarithms. Take $x = \frac{1}{99}$ and get $\ln\frac{100}{98} = \ln\frac{50}{49}$. Since we already have enough information to find $\ln 50$, it is now easy to find $\ln 49$, and half of that is $\ln 7$. Very clever. Euler shows off his method by finding logarithms of the integers from 1 to 10 to twenty-five decimal places.

At the end of Chapter VI there are six examples, five of which are "word problems." They are the only word problems in the entire *Introductio*, and there are no word problems at all in the two sequels, *Calculi differentialis* and *Calculi integralis*, so these are worth special attention.

Example 1 is not a word problem, but it has a word problem behind it. The problem is "to find the value of the power $2^{\frac{7}{12}}$." The solution is very simple by Eulerian standards, and the answer turns out to be 1.498307.

Euler doesn't mention it, but it is clear why he chose this problem. It is because the answer is so close to 1.5. This is important in the theory of music. If we vibrate two strings of equal density and tension, where one string is 1.5 times as long as the other, then the notes they sound will differ by what musicians call a "fifth." The scales built on this "perfect fifth" are called "Pythagorean," for reasons that are interesting, but too lengthy for this column. The notes on a Pythagorean scale are not quite evenly spaced. Euler, who studied the physics of music with great enthusiasm, [G] proposed an "even tempered" scale, for which the notes are spaced by making the ratio of the lengths of two consecutive strings be $\sqrt[12]{2}$. This example shows that the fifth note up on this scale differs from a Pythagorean fifth by only about one-tenth of one percent.

Euler's second problem reads "If the number of people who inhabit a certain province grows by one part in 30 each year, and initially the province has 100,000 people, find the number of inhabitants after 100 years." Euler uses the elementary properties of logarithms to find that the logarithm of the answer is $100 \log \frac{31}{30} + \log 100{,}000 = 6.4240439$, so the number of inhabitants will be 2,654,874.

Today this is a routine problem in exponential growth, but in 1748 it was one of the first hints that mathematics might be useful in the social sciences. At the time, Berlin was one of the world's few large cities to do a census. A careful census in 1747 showed the population of Berlin to be 107,224. [L] Other cities tried to estimate their populations using birth and death records, and such methods suggested that London had grown from about 100,000 to about a million over the previous 100 years. This problem, and the three that follow, give mathematical demonstrations that such ten-fold growth in a century is possible, even for large populations like that of Berlin.

Euler's next three examples are also population questions:

After the Flood, all people are descended from six people. If we suppose that after 200 years the population had grown to 1,000,000, then by what part must the population grow each year?

Euler is asking for a growth rate. Today we would probably express the growth rate as a percent. Euler sets up the problem to find x if

$$6\left(\frac{1+x}{x}\right)^{200} = 1000000.$$

He finds a rate of about 1/16 per year, that we would write as 6.25% per year.

The problem is to find the annual growth rate if the population doubles in a century. Here again, Euler expresses his growth rate as a fraction, 1/144, rather than the percentage, about 0.7%, as we probably would today.

The last population problem is to find how long it will take the population to double if the growth rate is 1/100 per year. Euler takes logarithms to get 231 years.

Euler's last word problem is a compound interest problem, and is kind of tricky. Somebody borrows 400,000 florins at an "annual rate of usury of 5 percent." (They didn't call it "interest" yet.) If he pays 25,000 florins each year, then how long will it take to repay the debt? Euler solves this by letting a be the 400,000 florins in the debt and b the 25,000 florins in each payment. Then he writes the debt after one year as $\frac{105}{100}a - b$, after two years as

$$\left(\frac{105}{100}\right)^2 a - \left(\frac{105}{100}\right) b - b.$$

Now he writes $n = \frac{105}{100}$ and x as the number of years, and gets the debt after x years as

$$n^x a - n^{x-1} b - n^{x-2} b - n^{x-3} b - \cdots - b.$$

Euler sees the geometric series in this and writes it as

$$n^x a - b\left(1 + n + n^2 + \cdots + n^{x-1}\right).$$

The sum of the finite geometric series inside the parentheses is

$$\frac{n^x - 1}{n - 1}.$$

Since we want to find the value of x that makes the debt equal to zero, we get the equation

$$n^x a = \frac{n^x b - b}{n - 1}.$$

After substituting the given values for n, a and b, we use logarithms to find that x is a little less than 33, so the debt is paid off after 33 years.

Today, this seems like a rather difficult but otherwise fairly routine problem in the mathematics of finance.

Textbooks have changed a good deal in the last 250 years. Euler's *Introductio* was a new kind of textbook in its day. Its presentation of logarithms is very similar to the one we use today, though the actual calculation of logarithms became obsolete with the advent of calculators. It is comforting to know, though, that Euler did not invent the "fake" word problem. His very few word problems were real problems that addressed actual issues that were important and interesting in the 1740s. It would be nice if modern authors could craft their products so carefully.

References

[E101] Euler, Leonhard, *Introductio in analysin infinitorum*, 2 vols., Bousquet, Lausanne, 1748, reprinted in the *Opera Omnia*, Series I vols. 8 and 9. English translation by John Blanton, Springer-Verlag, 1988 and 1990. Facsimile edition by Anastaltique, Brussels, 1967.

[G]   Greated, Clive, "Leonhard Euler," *The New Grove Dictionary of Music and Musicians*, 2ed.,Macmillan, New York, 2001, vol. 8, pp. 415–416.

[L]   Lewin, C. G., *Pensions and Insurance before 1800: A Social History*, Tuckwell Press, East Lothian, Scotland, 2003.

21

Roots by Recursion

(June 2005)

Euler's great work of 1748, *Introductio in analysin infinitorum*, is one of the world's truly great mathematics books. It was one of the first important books to be based on the concept of function, instead of the older idea of curves. It was also one of the first books specifically designed to help a student bridge the gap between algebra and calculus. As such, it, along with Maria Agnesi's (1718–1799) book *Instituzioni analitiche ad uso della gioventù italiana* that appeared the same year, should be considered to be the first precalculus books.

After writing the book, Euler had trouble finding a publisher. The academy in St. Petersburg that published many of his other books was suffering under the political turmoil in Russia at the time, so they did not have the resources to publish it. The Berlin Academy, where Euler was employed from 1741 to 1766, did not publish such books, so Euler found a publisher in Lausanne, Switzerland to produce the book. Euler was not able to journey to Lausanne to check the proofs of the book, so he asked his friend in Lausanne, Gabriel Cramer, to do it for him. This is the same Cramer of Cramer's rule in linear algebra and Cramer's paradox, the subject of our column from August 2004. Their correspondence about the production of the book provides some interesting insights into Euler's thoughts and the reception of his ideas in his own time.

The *Introductio* was published in two books, with Eneström numbers 101 and 102. Both have been translated by John Blanton and published in two volumes by Springer-Verlag in 1988 and 1990.

This month's column comes from Chapter 17 of Part I of the *Introductio*. The chapter is titled "On the use of recurrent series in finding the roots of equations."

In Chapter 13, "On recurrent series," Euler had described some results of DeMoivre on the expansion of rational functions into infinite series. In particular, using Euler's notation, if

$$\frac{a + bz + cz^2 + dz^3 + ez^4 + \text{etc.}}{1 - \alpha z - \beta z^2 - \gamma z^3 - \delta z^4 - \text{etc.}}$$

is a "proper" rational function, that is, the quotient of polynomials with the degree of the numerator less than the degree of the denominator, then it can be expanded into a power series

$$A + Bz + Cz^2 + Dz^3 + Ez^4 + Fz^5 + \text{etc.}$$

and after some complications in the initial terms, successive coefficients ... P, Q, R, S and T would satisfy a recurrence relation that Euler wrote as

$$T = \alpha S + \beta R + \gamma Q + \delta P + \text{etc.}$$

The initial complications are described by the following rules:

$$A = a$$
$$B = \alpha A + b$$
$$C = \alpha B + \beta A + c$$
$$D = \alpha C + \beta B + \gamma A + d\delta$$

These are not hard to check by multiplying the series together and matching terms, but we can make sure we know what we're doing by working through Euler's example: to expand

$$\frac{1-z}{1-z-2zz}$$

into a series.

If we set

$$\frac{1-z}{1-z-2zz} = A + Bz + Cz^2 + Dz^3 + Ez^4 + \text{etc.}$$

and multiply through by the $1 - z - 2zz$, we get

$$1 - z = (1 - z - 2zz)(A + Bz + Cz^2 + Dz^3 + Ez^4 + \text{etc.})$$

Matching constant terms we get

$$1 \cdot 1 = A \quad \text{so} \quad A = 1.$$

Matching linear terms we get

$$-z = 1 \cdot Bz - z \cdot A \quad \text{so} \quad B = 0.$$

Matching quadratic terms we get

$$0 = 1 \cdot Cz^2 - z \cdot Bz - 2zz \cdot A \quad \text{so} \quad C = B + 2A.$$

After that, the left-hand sides are all zero, and the coefficients on the right satisfy relations that proceed

$$D = C + 2B,$$
$$E = D + 2C,$$
$$F = E + 2D, \text{etc.}$$

The reader will want to continue this calculation and find that the first several coefficients are 1, 1, 3, 5, 11, 21 and 43.

A reader of the *Introductio* learned all of this in Chapter 13 and is ready to use it in Chapter 17. Let us proceed the way Euler does, and make some simplifying assumptions. Let's suppose we want to find a root of a cubic polynomial $1 - \alpha z - \beta z^2 - \gamma z^3$. Euler, following his usual pedagogy, starts with second degree polynomials, then does third degree, and then goes on to polynomials of arbitrary degree. We will do the derivation for third degree, but our example will be second degree.

Let us avoid a few complications and suppose also that the polynomial has three distinct real roots of different magnitudes, and, since the constant term is nonzero, none of the roots are zero.

Now, we know that the polynomial factors somehow as $(1 - pz)(1 - qz)(1 - rz)$, but if we knew exactly how it factors, we would know the roots are $1/p$, $1/q$ and $1/r$. Having said this, we know that the reciprocal of the polynomial can be written as a partial fraction:

$$\frac{1}{1 - \alpha z - \beta z^2 - \gamma z^3} = \frac{\mathfrak{A}}{1 - pz} + \frac{\mathfrak{B}}{1 - qz} + \frac{\mathfrak{C}}{1 - rz}$$

Euler has to resort to Fraktur characters because subscripts have not yet been invented. We do not need to know $\mathfrak{A}$, $\mathfrak{B}$, $\mathfrak{C}$ or p, q, r to know that the quotient has such a partial fraction representation.

Now, on the left, we can use the results from Chapter 13 to expand the quotient into a power series

$$A + Bz + Cz^2 + Dz^3 + Ez^4 + Fz^5 + \text{etc.}$$

where $A = 1$, and every coefficient after that is given by a simple recurrence relation involving α, β and γ. Meanwhile, each of the terms on the right can be expanded into geometric series:

$$\frac{\mathfrak{A}}{1 - pz} = \mathfrak{A} + \mathfrak{A}pz + \mathfrak{A}(pz)^2 + \mathfrak{A}(pz)^3 + \mathfrak{A}(pz)^4 + \text{etc.}$$

$$\frac{\mathfrak{B}}{1 - qz} = \mathfrak{B} + \mathfrak{B}qz + \mathfrak{B}(qz)^2 + \mathfrak{B}(qz)^3 + \mathfrak{B}(qz)^4 + \text{etc.}$$

$$\frac{\mathfrak{C}}{1 - rz} = \mathfrak{C} + \mathfrak{C}rz + \mathfrak{C}(rz)^2 + \mathfrak{C}(rz)^3 + \mathfrak{C}(rz)^4 + \text{etc.}$$

So, matching terms we get

$$A = \mathfrak{A} + \mathfrak{B} + \mathfrak{C}$$
$$B = \mathfrak{A}p + \mathfrak{B}q + \mathfrak{C}r$$
$$C = \mathfrak{A}p^2 + \mathfrak{B}q^2 + \mathfrak{C}r^2$$

and in general, for larger exponents,

$$M = \mathfrak{A}p^m + \mathfrak{B}q^m + \mathfrak{C}r^m$$
$$N = \mathfrak{A}p^{m+1} + \mathfrak{B}q^{m+1} + \mathfrak{C}r^{m+1}$$

Since we assumed that the roots are different magnitudes, one of p, q, r has the largest magnitude. Suppose it is p. This makes $1/p$ the smallest root. So, for large enough values of m, the term involving p dominates the other two terms, so

$$\frac{N}{M} = \frac{\mathfrak{A}p^{m+1} + \mathfrak{B}q^{m+1} + \mathfrak{C}r^{m+1}}{\mathfrak{A}p^m + \mathfrak{B}q^m + \mathfrak{C}r^m} \approx \frac{\mathfrak{A}p^{m+1}}{\mathfrak{A}p^m} = p$$

and, since p is the largest of the three coefficients in the factors, it is the smallest (in magnitude) of the three roots.

Let us try this with Euler's own first example: to find a root of $xx - 3x - 1 = 0$. Rather than consider $\frac{1}{1-3z-zz}$, Euler decides to look at $\frac{a+bz}{1-3z-zz}$, where a and b are whatever numbers make the sequence of coefficients begin $1, 2, \ldots$.

After that, the recurrence relation will give

$$C = 3B + A$$
$$D = 3C + B$$
$$\text{etc.}$$

The sequence thus generated is

$$1,\ 2,\ 7,\ 23,\ 76,\ 251,\ 829,\ 2738,\ \&c.$$

Taking the last pair of these, we get

$$p = \frac{829}{2738}$$
$$\frac{1}{p} = \frac{2738}{829}$$
$$= 3.3027744.$$

Finding the root by the quadratic formula gives

$$x = \frac{3 + \sqrt{13}}{2}$$
$$= 3.3027756.$$

The error is in the 6th decimal place.

Euler does a number of other examples, all of which are designed so that he has a way to check the results. Example IV is to find the root of $0 = 8y^3 - 24yy + 8y - 1$. The root is $1 + \sin 70°$. Euler finds that the sequence that begins $1, 1, 1 \ldots$ gives, as the ratio of its 8th and 9th terms, the correct answer to seven decimal places. Of course, most of Euler's examples are chosen so that they will work well.

Euler describes some enhancements. If, for example, we want to find the largest root of $f(z)$, we can substitute $z = \frac{1}{x}$, clear denominators, and find the smallest root. On the other hand, if we know that there is a root near $z = 2$, we can substitute $x = z - 2$ and the process will converge more quickly.

This method of Euler is seldom used today. It is considerably slower than Newton's method, though it does have some advantages, as the reader who is doing calculations by hand may already have discovered. Newton's method requires long division at every step. If the coefficients are simple, as they have been in every example, then Euler's method requires only one long division at the end. The reader is encouraged to work out some more examples and see if there are any other advantages or disadvantages.

References

[E101] Euler, Leonhard, *Introductio in analysin infinitorum*, 2 vols., Bousquet, Lausanne, 1748, reprinted in the *Opera Omnia*, Series I vols. 8 and 9. English translation by John Blanton, Springer-Verlag, 1988 and 1990. Facsimile edition by Anastaltique, Brussels, 1967.

22

Theorema Arithmeticum

(March 2005)

Euler's 1748 textbook, the *Introductio in analysin infinitorum*, was one of the most influential mathematics books of all time. John Blanton's excellent translation is available in many libraries and on the bookshelves of those few individuals who were lucky enough to get copies in the relatively brief time it was in print. Those people must to be holding on to their copies, as there seem to be almost none available on the used book market.

If you read the *Introductio*, you are likely to have different reactions to different parts of the book. When you read the section about partial fractions or the section about the definition of the trigonometric functions, you will feel very much at home. Euler's treatment is very similar to the way we present these topics today. This is because, for this topic, how Euler did it was adopted as the standard way to do it.

When you read about series, you may feel as if Euler is doing some things the hard way because he doesn't use calculus. This is a precalculus book, an *introduction* to the methods and material that will be used in calculus, so he does not use Taylor series or other calculus tools. It is surprising how much he is able to do without calculus.

Some other topics, like partitions and continued fractions, aren't seen so often any more, and it is exciting to see how much can be done by elementary means.

The *Introductio* provides a kind of foundation for much of Euler's career. Time and time again he finds a lemma in the *Introductio* that he needs in some later paper, or he writes a whole paper that begins with a topic from the *Introductio*.

This month's column, though, isn't about the *Introductio*. It is about something that would have fit well with some of the other material in the *Introductio*. Maybe it should have been there. When Euler needed this little result, it wasn't there, so he had to pause to prove it.

Euler wrote a massive text on calculus. His *Institutiones calculi differentialis*, E212, was published in 1755, seven years after the *Introductio*. More than ten years later, in 1769, E342, E366 and E385, his three volumes of the *Institutionum calculi integralis* came out. At more than 2500 pages, these four volumes outweigh even the most prolix of modern texts. Though Euler was seldom accused of being too brief, we should deflect some criticism of his verbiage; he does include both an extensive treatment of differential equations and a good bit of the calculus of variations under his umbrella of "calculus."

The whole calculus series is presented as a series of problems. Book I of the *Calculi integralis*, for example, has 173 problems, spread across two volumes. Each problem is given in rather general form, and with a general solution. Most solutions are followed by a number of corollaries, scholions or examples. Each problem, corollary or other part has a "paragraph" number, though most consist of more than one paragraph. Book I has 1275 such paragraphs.

Our example comes from near the end of Book I of the *Calculi integralis*, part of volume 2 of book I, so this is found in E366. It follows Problem 152 and is in paragraph 1169. At this point, Euler has been doing differential equations for over 300 pages. He comes upon a rather complicated problem (there isn't space to get into it here) that can be dramatically simplified using a clever partial fraction expansion.

Normally at this point, Euler would refer to the *Introductio* to find the lemma that solves the problem. This time, the result isn't there! So Euler pauses to give us:

Theorema Arithmeticum

Given numbers a, b, c, d, etc., if from each one is subtracted each other one and the following products are formed:

$$(a - b)(a - c)(a - d)(a - e) \text{ etc.} = \alpha$$
$$(b - a)(b - c)(b - d)(b - e) \text{ etc.} = \beta$$
$$(c - a)(c - b)(c - d)(c - e) \text{ etc.} = \gamma$$
$$(d - a)(d - b)(d - c)(d - e) \text{ etc.} = \delta$$

then it will always be that

$$\frac{1}{\alpha} + \frac{1}{\beta} + \frac{1}{\gamma} + \frac{1}{\delta} + \text{ etc.} = 0.$$

Euler overlooks the condition that the numbers a, b, c, d, etc., ought to be distinct, or else two of the products will be zero and the formula in the conclusion will be undefined.

If we have three numbers, a, b and c, then Euler is claiming that

$$\frac{1}{(a - b)(a - c)} + \frac{1}{(b - a)(b - c)} + \frac{1}{(c - a)(c - b)} = 0$$

With a bit of algebra, the reader who is careful with signs can easily verify this identity by using $(a - b)(a - c)(b - c)$ as a common denominator.

The case of four numbers, though, would require a common denominator with six factors, $(a - b)(a - c)(a - d)(b - c)(b - d)(c - d)$, and the algebra is considerably more cumbersome. In general, m numbers would require

$$\binom{m}{2} = \frac{m^2 - m}{2}$$

factors. This quickly moves from the awkward to the infeasible. There must be a better way.

The modern reader would probably rewrite Euler's claim using subscripts, sigmas and product symbols. Let the m numbers be a_1, a_2, a_3, ..., a_m, and the m products be given by

$$\alpha_i = \prod_{\substack{j=1 \\ j \neq i}}^{m} (a_i - a_j).$$

Then Euler claims that $\sum_{i=1}^{m} \frac{1}{\alpha_i} = 0$. Then, with a careful management of subscripts and symbols, it is probably possible to prove the result. It would probably not seem clever.

Euler though, did not have those tools, so he had to find a clever way.

Euler begins his proof with a step that makes the reader expect a proof by mathematical induction. That's not what he's doing, though. He supposes that the last of his m numbers is denoted by z, and that Z is a polynomial in z of degree less than $m - 1$. He forms the rational expression

$$\frac{Z}{(z-a)(z-b)(z-c)(z-d) \text{ etc.}}.$$

Note that the denominator here will be the last of the products Euler defined in the statement of his theorem, so that

$$\zeta = (z-a)(z-b)(z-c)(z-d) \text{ etc.}$$

Now, since he knows his *Introductio*, he decomposes this into its partial fractions:

$$\frac{A}{z-a} + \frac{B}{z-b} + \frac{C}{z-c} + \frac{D}{z-d} + \text{etc.} \tag{1}$$

We will need the negative of this expression, so Euler notes that its negative is

$$\frac{A}{a-z} + \frac{B}{b-z} + \frac{C}{c-z} + \frac{D}{d-z} + \text{etc.} \tag{2}$$

He is really only interested in the special case where $Z = z^n$, with n less than $m - 1$. For this particular Z, we can do just a little work and find the numbers A, B, C, etc. explicitly as

$$A = \frac{a^n}{(a-b)(a-c)(a-d) \text{ etc.}}$$

$$B = \frac{b^n}{(b-a)(b-c)(b-d) \text{ etc.}}$$

$$C = \frac{c^n}{(c-a)(c-b)(c-d) \text{ etc.}}$$

etc.

The last factors of these denominators are $(a - z)$, $(b - z)$, $(c - z)$, etc. Since Euler will be interested in the expansion of z^n/ζ, he will be using fractions for which the denominators involve $(z - a)$, $(z - b)$, $(z - c)$, etc., hence his remark that gave us equation (2).

With this groundwork set out, Euler is ready to look at the problem itself. Taking y to denote the penultimate term, the products given in the theorem are now:

$$(a - b)(a - c)(a - d)\cdots(a - z) = \alpha$$
$$(b - a)(b - c)(b - d)\cdots(b - z) = \beta$$
$$(c - a)(c - b)(c - d)\cdots(c - z) = \gamma$$
$$(d - a)(d - b)(d - c)\cdots(d - z) = \delta$$

etc.

$$(z - a)(z - b)(z - c)\cdots(z - y) = \zeta.$$

We notice that

$$\frac{z^n}{\zeta} = \frac{z^n}{(z - a)(z - b)(z - c)\cdots(z - y)}.$$

As we do the next few calculations, keep in mind that we just found a partial fraction expansion of this last expression. We see that

$$\frac{a^n}{\alpha} = \frac{a^n}{(a - b)(a - c)\cdots(a - y)(a - z)}$$

$$= \frac{a^n}{(a - b)(a - c)\cdots(a - y)} \cdot \frac{1}{(a - z)}$$

$$= A \cdot \frac{1}{a - z}$$

$$= \frac{-A}{z - a}.$$

Similarly for b^n/β, c^n/γ, etc. Now, putting this together with our partial fraction expansion, we can do the following calculation:

$$\frac{z^n}{\zeta} = \frac{z^n}{(z - a)(z - b)(z - c)\cdots(z - y)}$$

$$= \frac{A}{z - a} + \frac{B}{z - b} + \frac{C}{z - c} + \cdots + \frac{Y}{z - y}$$

$$= -\frac{a^n}{\alpha} - \frac{b^n}{\beta} - \frac{c^n}{\gamma} - \cdots - \frac{y^n}{\upsilon}$$

Now comes the punch line. From this last equation we get

$$\frac{a^n}{\alpha} + \frac{b^n}{\beta} + \frac{c^n}{\gamma} + \cdots + \frac{z^n}{\zeta} = 0.$$

Taking $n = 0$ gives the desired result.

There are probably many other ways to prove this result, but probably no other way has such an unexpected and surprise ending, and still uses only 18th century methods. Once again, Euler shows why he was the greatest of his century.

References

[E101] Euler, Leonhard, *Introductio in analysin infinitorum*, 2 vols., Bousquet, Lausanne, 1748, reprinted in the *Opera Omnia*, Series I vols. 8 and 9. English translation by John Blanton, Springer-Verlag, 1988 and 1990. Facsimile edition by Anastaltique, Brussels, 1967.

[E366] ——, *Institutionum calculi integralis, volumen secundum*, St. Petersburg, 1769, reprinted in *Opera Omnia* Series I vol. 12. Available online at `EulerArchive.org`.

Thanks to Rob Bradley for his help with this column.

23

A Mystery about the Law of Cosines

(December 2004)

The Law of Cosines has been in newspapers and magazines lately. Perhaps you have seen an advertisement that reads "Margaret needs to know what the heck $a^2 - 2ab \cos \theta + b^2$ is all about." They are trying to recruit people to teach high school mathematics. Those of us who recognize Margaret's formula as part of the Law of Cosines would make good candidates.

The day I first saw this advertisement, I had been reading parts of Leonhard Euler's *Calculi Integralis*, published in 1768. There, I found this same form in a very different context, and I thought it was mysterious.

Euler wrote a three-volume text on integral calculus. The volumes appeared in 1768, 1769 and 1770 and bear Eneström numbers 342, 366 and 385. Together with the two volumes of the *Introductio in analyisin infinitorum*, E102 and 101, published in 1748, and the *Calculi differentialis*, E212, 1755, and weighing in at over 2500 pages, they form the first really thorough set of calculus textbooks. They are often described as forming the basis for the modern calculus curriculum. This is something of an exaggeration, though. Much of the modern curriculum is missing from Euler. For example, Euler does no applications outside mathematics. There are no related rates problems or problems in physics. In fact, there are no exercises at all. On the other hand, Euler includes much that is not in most modern calculus courses. He does a lot of differential equations that we usually do in a separate course. Volume III ends with a long chapter on the calculus of variations, and Chapter 6 of the second part of Volume I covers a good deal about elliptic integrals, including the segment addition theorem. Both of these topics are very rare in the modern calculus curriculum.

Each volume has two or three "parts," and each "part" has about ten "chapters." A typical chapter consists of several "problems," each followed by a solution and several corollaries and scholions. Chapter 2 of the first part of Volume I is titled "De integratione formularum differentialium irrationalium," which translates as "On the integration of irrational differential formulas." It opens with:

Problem 6. Given a differential formula

$$dy = \frac{dx}{\sqrt{\alpha + \beta x + \gamma x x}},$$

to find its integral.

Euler's solution considers two cases. The first case is that the quadratic has two distinct real roots, and the second is that the quadratic is irreducible. Euler does not consider what might be a third case, that the quadratic has two equal roots, for in that case, taking the square root in the denominator reduces the problem to a much easier problem.

Euler's first case is not that important to the point we want to make in this column, so we will sketch it very briefly. He supposes that the quadratic factors into $(a+bx)(c+dx)$. Then he takes $z = \sqrt{(f + gx)/(a + bx)}$ and rewrites the quadratic as $(a + bx)^2 zz$. He shows how the integral involves logarithms if the signs of a and g are the same, and involves trigonometric functions if they have different signs.

What interests us is when he says that if the quadratic doesn't factor, then he can write it as

$$dy = \frac{dx}{\sqrt{aa - 2abx \cos \zeta + bbxx}}$$

If we take $x = 1$, and substitute one Greek letter for another, then the expression inside the radical in the denominator is part of that Law of Cosines that the advertisement says Margaret needs to know! Who would have thought that irreducible quadratics would have anything to do with the Law of Cosines?

Is this quadratic really irreducible? Its discriminant is $a^2 b^2 (\cos^2 \zeta - 1)$, and that is never positive. When $\cos \zeta = 1$, the discriminant is zero and there is a double root, and we have already excluded that case. So, indeed, it is safe to say the quadratic really is irreducible.

Other questions remain. Can any irreducible quadratic be put in this form? What is the significance of the angle ζ? What are the roots of such a form? Why would Euler use such a form?

Readers who have pencils are encouraged to investigate these questions a bit before reading any farther.

There are a number of ways to approach this challenge, each with its own beautiful aspects. I posed this at dinner at a recent MAA Section meeting, and no two people at the table of six did it the same way. My favorite involved completing the square and an application of Euler's formula, $e^{i\theta} = \cos \theta + i \sin \theta$.

Rather than deprive the reader of the pleasure of discovering such pleasant derivations, I'll describe a less elegant approach. Suppose we are given the roots of an irreducible quadratic in polar form, say $(r, \pm\theta)$. Then we can write the roots in Cartesian form, as $s = r \cos \theta + i \sin \theta$ and $t = r \cos \theta - i \sin \theta$. When we expand $(x - s)(x - t)$, we get $x^2 - 2r \cos \theta + r^2$. From this, it is easy to see that any irreducible quadratic polynomial can indeed be put into Euler's form. In fact, just as it is easy to see that the roots are s and t when we write it in the form $a(x - s)(x - t)$, it is easy to see that, in polar form (and when a and b have the same sign), the roots of

$$aa - 2abx \cos \zeta + bbxx$$

are $\left(\frac{a}{b}, \pm\zeta\right)$. (The case of mixed signs is only slightly irksome.)

It might seem that we've solved most of the mystery; why is the form

$$aa - 2abx \cos \zeta + bbxx$$

irreducible? But in finding this answer, we have made a serious error in historical analysis. We have represented complex numbers in polar form, using a radius and an angle. This idea is usually said to have originated in a 1797 paper by Caspar Wessel [N, p. 48], so Euler, in 1768, should not have been able to use it.

So, we are left with an even more perplexing mystery. Euler uses an idea that we understand because we know about the polar form of a complex number. Euler had no such notion, yet he uses the idea as if it were natural and well known in its time. How did Euler know and understand that all irreducible quadratics could be written as $aa - 2abx \cos \zeta + bbxx$?

I don't know. And I also don't know how it is connected to the Law of Cosines. Maybe our young friend Margaret will be the one to figure it out.

References

[E342] Euler, Leonhard, *Institutionum calculi integralis volumen primum*, St. Petersburg, 1768, reprinted in *Opera Omnia* Series I vol. 11. Available on line at `EulerArchive.org`.

[N]   Nahin, Paul, *An Imaginary Tale: The Story of* $\sqrt{-1}$, Princeton Univ. Press, 1998.

24

A Memorable Example of False Induction

(August 2005)

Euler wrote about 800 books and papers. An exact number is hard to define. The "official" number of entries in Eneström's index is 866, but that includes a number of letters and unfinished manuscripts that Euler never expected to be published. Euler probably intended to finish some of the manuscripts, but others he had probably abandoned as dead-ends. Moreover, though most of Euler's letters were simple communications, some were more like "open letters," intended to be shared widely, so they were more like publications than private communication. Taking all of this into account, an estimate of "about 800" publications seems quite reasonable.

Euler wrote his first article in 1725, and it was published in 1726. He died in 1783, but papers intended for publication continued to appear until 1862, 79 years after his death. Below, we give a graph and a table describing the decades that Euler wrote 810 of his books and articles.

Of the 810 books and articles in this database, the Editors of Euler's *Opera Omnia* classified 81 of them, exactly 10%, as being about series, and so published them in volumes

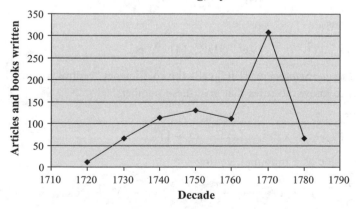

Euler's writing, by decade

Decade	Works
1720	11
1730	67
1740	114
1750	131
1760	111
1770	309
1780	67

14, 15 and 16 of Series I. The Editors used what some might think is an expanded definition of "series" that also includes infinite products and continued fractions. The timing of Euler's work in series has a somewhat different shape than his work as a whole, as seen in the graphics below.

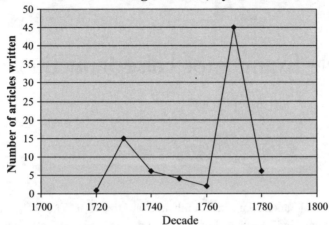

Euler's writing on series, by decade

Decade	Series papers
1720	1
1730	15
1740	6
1750	4
1760	2
1770	45
1780	6

Euler's interest in series seems to be declining through the heart of his career, in the 1740s and 1750s, to the point where he wrote only two papers on the subject in the whole of the 1760s. One of those papers was on properties of the Bernoulli numbers, and the other, the one we discuss here, on properties of a particular series.

This column's paper is E326, written in 1763 and titled *Observationes analyticae*, or "Analytical observations." Euler plans to sum the middle terms of powers of quadratics, starting with the very simple quadratic, $1 + x + xx$. He begins by listing the powers of $1 + x + xx$:

$$1$$
$$1 + \quad x \quad + xx$$
$$1 + 2x + \mathbf{3x^2} + 2x3 + x4$$
$$1 + 3x + 6x2 + \mathbf{7x^3} + 6x4 + 3x5 + x6$$
$$\text{etc.}$$

Now we look at the terms

$$1, \ 1x, \ 3x^2, \ 7x^3, \ 19x^4, \ 51x^5, \ 141x^6, \ \text{etc.}$$

The *Encyclopedia of Integer Sequences*, [EIC] calls the coefficients "central trinomial coefficients." Euler wants to know the rules that give these numbers.

He begins by rewriting

$$(1 + x + xx)^n = \left(x(1 + x) + 1\right)^n.$$

He expands the right-hand side as a binomial, getting

$$x^n(1 + x)^n + \frac{n}{1}x^{n-1}(1 + x)^{n-1} + \frac{n(n-1)}{1 \cdot 2}x^{n-2}(1 + x)^{n-2} + \text{etc.}$$

This is just the Binomial Theorem. It would look more familiar if Euler had written this paper just a few years later, after he introduced an almost modern notation for binomial coefficients, writing $\left(\frac{n}{k}\right)$ where we would usually write $\binom{n}{k}$.

This last expression still contains binomials, so Euler steadfastly expands it again and combines like terms to find that the coefficient of x^n is

$$1 + \frac{n(n-1)}{1 \cdot 1} + \frac{n(n-1)(n-2)(n-3)}{1 \cdot 2 \cdot 1 \cdot 2} + \frac{n(n-1)(n-2)(n-3)(n-4)(n-5)}{1 \cdot 2 \cdot 3 \cdot 1 \cdot 2 \cdot 3} + \text{etc.}$$

Armed with this formula, Euler calculates the first 12 terms of his sequence. If he had access to the online *Encyclopedia of Integer Sequences*, then in just a few moments he could have found over 20 terms:

1, 1, 3, 7, 19, 51, 141, 393, 1107, 3139, 8953, 25653, 73789, 212941, 616227, 1787607, 5196627, 15134931, 44152809, 128996853, 377379369, 1105350729, 3241135527, 9513228123, 27948336381, 82176836301, 241813226151,...

Having found a direct formula for the central trinomial coefficients, and listing the first twelve coefficients, up to 73789, Euler begins a section mysteriously titled:

EXEMPLUM MEMORABILE INDUCTIONIS FALLACIS

Formulas are sometimes cumbersome, and this formula is particularly so. True to form, Euler sets out to find a recursive formula for these numbers. He writes his sequence in one row, the triple of the sequence, offset by one position, in the second row, and subtracts the first row from the second. It looks like this:

1	1	3	7	19	51	141	393	1107	3139	etc.
	3	3	9	21	57	153	423	1179	3321	etc.
	2	0	2	2	6	12	30	72	182	etc.

Now Euler notices (*non sine ratione evenire videtur*, "not without thought it is seen to turn out") that all the numbers in this last row are the double of triangular numbers, and so have the form $mm + m$, for various values of m. Some people used to call these products of consecutive integers of the form $m(m + 1)$ *oblong* or *Pronic* numbers.

So, what values of m give these particular values if $mm + m$? Euler calls these values of m the *indices*, and the indices go

1, 0, 1, 1, 2, 3, 5, 8, 13, etc.

This is the Fibonacci sequence, starting just a little bit early, with first two terms 1 and 0, rather than the more familiar starting values 1 and 1. Actually, this apparently wasn't *called* the Fibonacci sequence until the late 1800s, but that wouldn't keep Euler from knowing a lot about the sequence. In particular, he knows from his work on difference equations and generating functions that the nth term of this sequence is given by the formula

$$\frac{1}{\sqrt{5}} \left(\frac{1 + \sqrt{5}}{2}\right)^{n-2} - \frac{1}{\sqrt{5}} \left(\frac{1 - \sqrt{5}}{2}\right)^{n-2}.$$

Let us do something Euler couldn't do, because subscripted sequences hadn't been invented yet, and denote these values by f_n.

From this, Euler can deduce a recursive formula. If we write the sums of the central trinomial terms as

$$1 + x + 3x^2 + 7x^3 + 19x^4 + \cdots + Px^n + Qx^{n+1} + \text{etc.}$$

then for the data in this table $3P - Q = (f_n + 1)f_n$.

Euler also derives a direct formula for P, and a second recursive formula that is homogeneous (i. e., one that does not involve n.)

With these, we can find central trinomial coefficients quickly and easily, and we get

$$1, \ 1, \ 3, \ 7, \ 19, \ 51, \ 141, \ 393, \ 1107, \ 3139, \ 8955, \ 25675, \ 73945, \ \text{etc.}$$

But wait a minute! This isn't the same sequence we started with! It is the same for the first nine terms, up to 3139, but for the tenth term, this has 8955 where there should be an 8953, and after that the differences become even larger. Now we see the meaning of the title of this section, which translates as "A notable example of false induction." He had warned us. There really are two different sequences, each defined by reasonable and interesting patterns that agree for the first nine terms, and then become different.

Euler still has an article to finish, but nothing else this interesting happens. He finds the correct recursive relation directly from the formulas (so it is correct): if P, Q and R are consecutive coefficients, then the nth term is given by the relation

$$R = Q + \frac{n+1}{n+2}(Q + 3P).$$

That done, he spends the rest of the article investigating the central coefficients of powers of general quadratics of the form $a + bx + cxx$.

In 1753, Euler had written an E256, *Specimen de usu observationum in mathesi pura*, "Example of the use of observation in pure mathematics." It was an article about number theory, showing how experiments on integers of the form $a^2 + 2b^2$ led him to observe that such forms are closed under multiplication. This told him what to try to prove, and soon led to a proof of that and several related results.

Ten years later, he gives us a graphic illustration of the limits of observation, and that it shows mathematicians what might be true, not necessarily what is true.

References

[E256] Euler, Leonhard, Specimen de usu observationum in mathesi pura, *Novi commentarii academiae scientiarum Petropolitanae*, 6 (1756/7) 1761, pp. 185–230. Reprinted in *Opera Omnia* Series I vol. 2 pp. 459–493. Available online at EulerArchive.org.

[E326] ——, Observationes analyticae, *Novi commentarii academiae scientiarum Petropolitanae*, 11 (1765) 1767, pp. 124–143, reprinted in *Opera Omnia* Series I vol. 15 pp. 50–69. Available online at EulerArchive.org.

[EIS] Encyclopedia of Integer Sequences, on line at research.att.com/ njas/sequences/, consulted July 21, 2005.

[W] Weisstein, Eric W. "Pronic Number." From MathWorld—A Wolfram Web Resource. mathworld.wolfram.com/PronicNumber.html.

25

Foundations of Calculus

(September 2006)

As we begin a new academic year, many of us are introducing another generation of students to the magic of calculus. As always, those of us who teach calculus are asking ourselves again, "What is the best way to begin calculus?" More specifically, "How do we start to teach students what a derivative is?" Some of us will start with slopes and others choose limits. Among those who begin with limits, some will use epsilons and deltas and others will use a more intuitive approach to the algebra of limits. A few might use non-standard analysis, as rigorously presented in the wonderful book [K] by Jerome Keisler. Newton began with "fluxions," while Leibniz used differentials and a "differential triangle." Regular readers of this column, though, know to ask "How did Euler do it?"

Almost as soon as it was invented (or, if you prefer, discovered) people began arguing about its foundations. Leibniz looked for an algebraic basis for calculus, while Newton argued in favor of geometric foundations. The controversy continued for more than a hundred years, with key contributions from Berkeley and Lagrange, until most of the issues were finally resolved in the time of Cauchy, Riemann and Weierstrass.

Euler published his differential calculus book, *Institutiones calculi differentialis*, [E212], in 1755. The book has two parts. Euler describes the first part, nine chapters, 278 pages in the original, as "containing a complete explanation of this calculus." John Blanton translated this part of the book into English in 2000, and most of the quotations used in this column are from John Blanton's edition. The second part of the book, 18 chapters, 602 pages, "contains the use of this calculus in finite analysis and in the doctrine of series." A translation of this part of the book has not yet been published, though there are rumors that people are working on it.

When Euler sat down to write the *Calculi differentialis*, as it is commonly called, he had to decide how to explain the foundations of calculus and the reasons calculus "works." In his preface he writes (in John Blanton's translation, p. vii.):

> [D]ifferential calculus ... is a method for determining the ratio of the vanishing increments that any functions take on when the variable, of which they are functions, is given a vanishing increment.

This echoes with Newtonian sentiments. "Vanishing increments" sound like Newton's "evanescent quantities," and are open to Berkeley's sarcastic barbs, calling infinitesimals "ghosts of departed quantities." Euler, who learned his calculus from Johann Bernoulli, a follower of Leibniz, understands these criticisms, and in the very next paragraph he writes

> [D]ifferential calculus is concerned not so much with vanishing increments, which indeed are nothing, but with the ratio and mutual proportion. Since these ratios are expressed as finite quantities, we must think of calculus as being concerned with finite quantities.

Later (p. viii) he writes

> To many who have discussed the rules of differential calculus, it has seemed that there is a distinction between absolutely nothing and a special order of quantities infinitely small, which do not quite vanish completely but retain a certain quantity that is indeed less than any assignable quantity.

Euler seems to want it both ways. He wants to use infinite numbers, usually denoted i or n, as well as infinitesimals (he calls them "infinitely small,") usually denoted ω. He wants to take their ratios, add, subtract and multiply them as if they matter, and then throw them away when it suits his purposes. It is exactly the behavior that Berkeley was trying to discourage and that Cauchy and Weierstrass eventually repaired.

Now that we've seen these philosophical underpinnings, let's look at how Euler teaches us calculus.

Euler's Chapter 1 is "On finite differences" (*De differentiis finitis*). Euler gives us a variable quantity x, and an increment ω. For now, ω is assumed to be finite. He asks how substituting $x + \omega$ for x in a function "transforms" that function, and gives us the example

$$\frac{a+x}{a^2+x^2} \quad \text{is transformed into} \quad \frac{a+x+\omega}{a^2+x^2+2x\omega+\omega^2}.$$

The value x and its increment ω give an arithmetic sequence, x, $x + \omega$, $x + 2\omega$, $x + 3\omega$, etc., and these, in turn, transform a function y into a sequence of values that he denotes y, y^I, y^II, y^III, etc.

With this notation in place, he is ready to describe the first, and higher differences of the function. This is apparently the first time the symbol Δ was used for this purpose. The image on the facing page, from The Euler Archive, shows Euler's definition of the higher differences.

Next we get some of the elementary properties of higher differences, starting with their values in terms of the values of y.

$$\Delta\Delta y = y^\mathrm{II} - 2y^\mathrm{I} + y$$
$$\Delta\Delta y^\mathrm{I} = y^\mathrm{III} - 2y^\mathrm{II} + y^\mathrm{I}$$
$$\Delta^3 y = y^\mathrm{III} - 3y^\mathrm{II} + 3y^\mathrm{I} - y.$$

and so on up to 5th differences. He follows these with rules for sums and products:

If $y = p + q$ then $\Delta y = \Delta p + \Delta q$, and
If $y = pq$ then $\Delta y = p\Delta q + q\Delta p$.

PROGRESSIO ARITHMETICA.

$$x; \; x+\omega; \; x+2\omega; \; x+3\omega; \; x+4\omega; \; x+5\omega; \; \&c.$$

VALORES FUNCTIONIS.

$$y; \; y^{\text{I}} \quad ; \; y^{\text{II}} \quad ; \; y^{\text{III}} \quad ; \; y^{\text{IV}} \quad ; \; y^{\text{V}} \quad ; \; \&c.$$

DIFFERENTIAE PRIMAE.

$$\Delta y; \; \Delta y^{\text{I}}; \; \Delta y^{\text{II}}; \; \Delta y^{\text{III}}; \; \Delta y^{\text{IV}}; \quad \&c.$$

DIFF. II. $\Delta\Delta y; \; \Delta\Delta y^{\text{I}}; \; \Delta\Delta y^{\text{II}}; \; \Delta\Delta y^{\text{III}}; \quad \&c.$

DIFF. III. $\Delta^3 y \; ; \; \Delta^3 y^{\text{I}}; \; \Delta^3 y^{\text{II}}; \quad \&c.$

DIFF. IV. $\Delta^4 y \; ; \; \Delta^4 y^{\text{I}} \; ; \; \&c.$

DIFF. V. $\Delta^5 y \; ; \; \&c.$
          $\&c.$

Note that in the product rule, Euler omits the term $\Delta p \Delta q$.

By the end of this 24-page chapter, Euler has taught us to find differences of sums, products and radicals involving polynomials, sines, cosines, logarithms and radicals, as well as inverse differences, finding a function that has a given first difference.

In Chapter 2 Euler uses differences to study what he calls "series." We would call them "sequences." Probably his most interesting result in the chapter is the discrete Taylor series. Consider a sequence $a, a^{\text{I}}, a^{\text{II}}$, etc., with indices 1, 2, 3, 4, etc. Let the first, second, third, etc. differences be b, c, d, etc. Then the "general term" of index x is given by

$$a + \frac{(x-1)}{1}b + \frac{(x-1)(x-2)}{1 \cdot 2}c + \frac{(x-1)(x-2)(x-3)}{1 \cdot 2 \cdot 3}d + \text{etc.}$$

In Euler's chapter 3, "On the Infinite and the Infinitely Small," he returns to the philosophical underpinnings of calculus. The chapter is 30 pages long, and is mostly a fascinating philosophy-laced essay on the nature of infinites and infinitesimals in the real world. In the 18th century, real numbers were not the free-standing axiomatized objects they became late in the 19th century (thanks to Dedekind and his cuts). The properties of the real numbers were expected to reflect the properties of the real world they describe. Hence, the debate between Newton and Leibniz about whether real world objects are infinitely and continuously divisible (Newton), or composed of indivisible ultimate particles (that Leibniz called *monads*) was also a dispute over the nature of the real numbers. That, in turn, became a dispute about the nature and foundations of calculus. Though their public argument was mostly about who first discovered calculus, each also believed that the other's version of calculus was based on false foundations.

After careful consideration, Euler eventually sides mostly with Newton and accepts the "reality" of the infinite and infinitely small. The difficult part of this is that, if dx is an infinitely small quantity, then

Since the symbol ∞ stands for an infinitely large quantity, we have the equation

$$\frac{a}{dx} = \infty.$$

The truth of this is clear also when we invert:

$$\frac{a}{\infty} = dx = 0.$$

So, Euler is stuck with the paradox that the quantity dx is, in some sense, both zero and not zero. He cannot resolve this paradox, so he has to figure out a way to avoid it.

To do this, he begins to use his results on differences from the first two chapters. When he takes a sequence with an infinitely small quantity dx as its difference, then considers second differences, he is forced to conclude that "a/dx^2 is a quantity infinitely greater than a/dx," and similarly for higher differences. "We have, therefore, an infinity of grades of infinity, of which each is infinitely greater than its predecessor."

Essentially, he introduces some rules for the use of infinite and infinitesimal quantities, roughly equivalent to our techniques for manipulating limits. A quantity like $a/dx : A/dx$, where a and A are finite quantities (i.e., neither infinite nor infinitesimal) should not be resolved by first dividing by dx, for that leads to ∞/∞. This cannot be evaluated because, as he noted above, there are many different sizes of infinity, and this expression doesn't tell us which size we have. Instead, the quantity $a/dx : A/dx$ should first be reduced to a/A, a ratio of finite quantities.

Euler tells us that also "it is possible not only for the product of an infinitely large quantity and an infinitely small quantity to produce a finite quantity, ... but also that a product of this kind can also be either infinitely large or infinitely small."

The last part of chapter 3 deals with issues of convergence. It could be the basis of some future column.

Now that Euler believes he has convinced us of the logical integrity of his foundations, he returns to his calculations with series and differences. He reminds us that, given a variable x, a quantity y that depends on x, and an increment $\Delta x = \omega$, then y has a difference of the form

$$\Delta y = y^{1} - y = P\omega + Q\omega^2 + R\omega^3 + S\omega^4 + \text{etc.}$$

Taking $\omega = dx$, we get that $dy = P\,dx$, or, as we would write it today, $\frac{dy}{dx} = P$. Note that Euler and his contemporaries did calculus with differentials, dy and dx, and not with derivatives, $\frac{dy}{dx}$. Euler explains how the higher coefficients, Q, R, S, etc., are related to higher differences, and he's ready to go with the rules of calculus.

Before he goes, though, he makes a remark to appease his Leibnizian friends by criticizing Newton's fluxion notation. Newton would write $\dot{y}$, $\ddot{y}$ or $\dddot{y}$ where Euler would write P, Q or R, or maybe dy, d^2y or d^3y. Euler writes that Newton's notation "cannot be criticized if the number of dots is small.... On the other hand, if many dots are required, much confusion and even more inconvenience may be the result." As an example, he gives (and here, for the first time, we don't follow Blanton's translation) "The tenth fluxion, though would be very inconveniently represented by $\overset{\vdots\vdots}{\dot{y}}$, and our notation of $d^{10}y$ is much easier to understand."

With this, Euler sets out to give the usual rules of differential calculus, of course using differentials instead of derivatives and, in part one, omitting all applications. This column is about Euler's foundations of calculus, so we will leave out most of the content, for now. One of Euler's examples, though, is particularly elegant, from chapter 6, "On the Differentiation of Transcendental Functions:"

If $y = e^{e^{e^x}}$, then $dy = e^{e^{e^x}} e^{e^x} e^x dx$.

It is visually striking if you write it on a blackboard and use a longer string of e's.

Euler is sometimes criticized by modern mathematicians for what seems like a reckless use of infinite and infinitesimal numbers in his calculations, and for ignoring the foundations of calculus. What he writes in the *Calculi differentialis*, though, makes it clear that he was very aware of the issues involved, and that he tried hard to resolve them. In fact, he himself believed that he had indeed put calculus on a solid philosophical foundation.

Two generations after Euler, though, the way we build the foundations of mathematics changed, and a philosophical basis was no longer accepted. The age of Cauchy and Weierstrass sought less geometric and more axiomatic foundations, and Euler's approach was discarded as insufficiently rigorous.

In the 20th century, though, Abraham Robinson developed non-standard analysis, and showed how Euler's techniques could be made rigorous. Jerome Keisler, [K] in turn, used Robinson's constructions to write a modern calculus text. It took 220 years, but Euler's *Calculi differentialis* was eventually shown to have rigorous foundations. Now, if we want to, we can do it the way Euler did it.

References

[E212] Euler, Leonhard, *Insitutiones calculi differentialis cum ejus usu in analysi finitorum ac doctrina serierum*, St. Petersburg, 1755. Reprinted in *Opera Omnia*, Series I vol. 10. English translation of chapters 1 to 9 by John Blanton, Springer, New York, 2000. Available online at EulerArchive.org.

[K]   Keisler, H. Jerome, *Elementary Calculus*, Prindle, Weber & Schmidt, Boston, 1976. The entire book is available (Free! under a Creative Commons License) at math.wisc.edu/~keisler/calc.html.

26

Wallis's Formula

(November 2004)

Besides everything else he did, Euler was the best mathematics textbook writer of his age, with a line of texts that extended from arithmetic to advanced calculus, and a popular book on general science as well. We summarize his textbook output below:

1738   *Rechenkunst*, Arithmetic, in German and Russian, 277 pages

1744   *Methodus inveniendi*, Calculus of variations, Latin, 322 pages

1748   *Introductio in analysin infinitorum*, Precalculus, Latin, 2 vols., 320 + 398 pages

1755   *Calculi differentialis*, Differential calculus, Latin, 880 pages

1768   *Lettres à une Princesse d'Allemagne*, Letters to a German princess, general science, French, 3 vols., 314 + 340 + 404 pages

1768   *Calculi integralis*, Integral calculus, Latin, 3 vols., 542 + 526 + 639 pages

1770   *Vollständige Anleitung zur Algebra*, German, 2 vols., 356 + 532 pages

That is quite a curriculum, and, at over 5800 pages and thirteen volumes quite a bookshelf. The calculus alone, differential and integral, is over 2500 pages, about twice the length of the larger modern texts, and Euler didn't include exercises. To be fair, it includes a lot of differential equations, but much of the material on series was covered in the precalculus text, the *Introductio*.

This month, celebrating the beginning of the second year of this column, we will look at a couple of paragraphs about infinite products from the middle of the first volume of the *Calculi integralis*, in a section titled "De evolutione integralium per producta infinita," or "On the expansion of integrals by infinite products."

The three volumes of the *Calculi integralis* are divided into parts, which are, in turn, divided into sections, then chapters and finally paragraphs. The first two volumes number 1275 paragraphs and 173 problems. The paragraph numbers and problem numbers start again at 1 for the third volume.

The *Calculi integralis* is, for the most part, organized as a series of problems and their solutions and generalizations. A "problem" is likely to have three or four corollaries.

The section "De evolutione integralium per producta infinita" begins with Problem 43 and paragraph 356 (both counting from the beginning of the book):

356. To expand the value of this integral

$$\int \frac{dx}{\sqrt{1-xx}}$$

in the case $x = 1$.

To a modern reader, this isn't very clear. It looks like an indefinite integral, but this is how Euler wrote definite integrals. He means us to take the particular antiderivative that is zero at the left-hand endpoint, in this case assumed to be zero, and then to evaluate this antiderivative at $x = 1$. So, we would write Euler's integral as

$$\int_0^1 \frac{dx}{\sqrt{1-x^2}}.$$

It is easy to see that this is an arcsine and it has value $\frac{\pi}{2}$. That's not the point, though.

For the preceding several sections, Euler has been doing gymnastics with integration by parts, so it fits in very naturally to use integration by parts here, not on the specific

$$\int \frac{dx}{\sqrt{1-xx}},$$

but on the more general

$$\int \frac{x^{m-1}dx}{\sqrt{1-xx}}.$$

He gets

$$\int \frac{x^{m-1}dx}{\sqrt{1-xx}} = \frac{m+1}{m} \int \frac{x^{m+1}dx}{\sqrt{1-xx}}.$$

We are still doing definite integrals, and the leading term outside the integral evaluates to zero at both endpoints, so that term disappears.

Applying this repeatedly, starting with the case $m = 1$, so $m - 1 = 0$ gives

$$\int \frac{dx}{\sqrt{1-xx}} = \frac{2}{1} \int \frac{xx\,dx}{\sqrt{1-xx}}$$

$$= \frac{2 \cdot 4}{1 \cdot 3} \int \frac{x^4\,dx}{\sqrt{1-xx}}$$

$$= \frac{2 \cdot 4 \cdot 6}{1 \cdot 3 \cdot 5} \int \frac{x^6\,dx}{\sqrt{1-xx}}.$$

Now, Euler makes a rather suspicious step, taking i to be an infinite number, extends this to infinity, and writes the infinite product

$$\int \frac{dx}{\sqrt{1-xx}} = \frac{2 \cdot 4 \cdot 6 \cdot 8 \cdots 2i}{1 \cdot 3 \cdot 5 \cdot 7 \cdots (2i-1)} \int \frac{x^{2i}\,dx}{\sqrt{1-xx}}.$$

Now, unexpectedly, we do the same thing with $\int \frac{x\,dx}{\sqrt{1-xx}}$, another definite integral which is known to be equal to 1. We get another equally dubious infinite product

$$\int \frac{x\,dx}{\sqrt{1-xx}} = \frac{3\cdot5\cdot7\cdot9\cdots(2i+1)}{2\cdot4\cdot6\cdot8\cdots2i} \int \frac{x^{2i+1}\,dx}{\sqrt{1-xx}}.$$

These products are a bit awkward, since the fraction part diverges to infinity, while the integral part goes to zero. Nineteenth century analysis taught us that such objects ought to be treated carefully. However, one of the principal motivations for that analysis was to understand why sometimes, as in the present case, manipulations that used infinite numbers would work, and lead to correct and consistent results, but that sometimes they would lead to contradictions. However, those are not the issues of 1768 when Euler published the *Calculi integralis*.

Euler asks us to observe that, if i is an infinite number, then these last factors of the two infinite products,

$$\int \frac{x^{2i}\,dx}{\sqrt{1-xx}} \quad \text{and} \quad \int \frac{x^{2i+1}\,dx}{\sqrt{1-xx}}$$

will be equal. This allows Euler to know the ratios between the two infinite products. He leads us to the ratio as follows:

Let

$$\frac{2\cdot4\cdot6\cdot8\cdots2i}{1\cdot3\cdot5\cdot7\cdots(2i-1)} = M \quad \text{and} \quad \frac{3\cdot5\cdot7\cdot9\cdots(2i+1)}{2\cdot4\cdot6\cdot8\cdots2i} = N.$$

Then, in the old ":" notation for ratios,

$$\int \frac{dx}{\sqrt{1-xx}} : \int \frac{x\,dx}{\sqrt{1-xx}} = M : N = \frac{M}{N} : 1$$

We know that the ratio of the two integrals is $\frac{\pi}{2}$, so the ratio of M to N ought to be the same thing. Writing that ratio as an infinite product, we get

$$\frac{\pi}{2} = \frac{M}{N} = \frac{2\cdot2}{1\cdot3}\cdot\frac{4\cdot4}{3\cdot5}\cdot\frac{6\cdot6}{5\cdot7}\cdot\frac{8\cdot8}{7\cdot9}\cdot\text{etc.}$$

This is the well-known Wallis Formula, first discovered by John Wallis (1616–1703) [OC]. Wallis was a part-time cryptographer for both sides of the religious wars in England, and Savilian Professor of Geometry at Oxford. He published this result in his most famous work, the book *Arithmetica infinitorum*, published in 1656, ten years before Newton discovered calculus, and almost 30 years before Leibniz published his results. In its time, Wallis's infinite product was so remarkable that some important mathematicians, including the great Christiaan Huygens, simply didn't believe it.

Euler had used Wallis's formula before, in 1729, when he discovered what we now call the Gamma function. There he reduced the Gamma function to an infinite product and compared that infinite product to Wallis's formula to find that $\Gamma(\frac{1}{2}) = \sqrt{\pi}$. Now, almost 40 years later and over a hundred years after Wallis, Euler uses Wallis's formula again to check that his bold calculations with ratios of infinite quantities are working.

References

[OC]   O'Connor, J. J., and E. F. Robertson, "John Wallis," The MacTutor History of Mathematics archive, www-gap.dcs.st-and.ac.uk/~history/Mathematicians/Wallis.html, February 2002.

[E242] Euler, Leonhard, *Institutionum calculi integralis*, vol. 1, St. Petersburg, 1768, reprinted in *Opera Omnia*, Series I vol. 11.

27

Arc Length of an Ellipse

(October 2004)

It is remarkable that the constant, π, that relates the radius to the circumference of a circle in the familiar formula $C = 2\pi r$ is the same constant that relates the radius to the area in the formula $A = \pi r^2$. This is a special property of circles. Ellipses, despite their similarity to circles, are quite different. There is no easy relationship between the circumference and the area of an ellipse.

On the one hand, if the two semi-radii of an ellipse are a and b, then the area of the ellipse is given by $A = \pi ab$. The constant π is the same constant that works for circles. The area of a circle is a special case of this. On the other hand, arc length on an ellipse is a deep and considerably more difficult question. As we will see, the arc length is given either by a hard integral or by a rather formidable series. Early work was done by the Italian mathematician Giulio Fagnano and the Swede Samuel Klingenstierna, but we will follow Euler's version.

Euler worked throughout his life on integrals involving the arc length of the ellipse. We will look at his earliest efforts, a paper written in 1732, published in 1738, in which he found a series for the arc length of a quarter of an ellipse. The result is part of a paper [E28] titled *Specimen de constructione aequationum differentialium sine indeterminatarum separatione*, or, in English, "Example of the construction of a differential equation without the separation of variables." As the title suggests, the arc length of the ellipse arises as Euler is pursuing a problem in differential equations.

As always, we begin with notation. In Figure 1, arc BMA is a quarter of an ellipse, and other parts are defined as follows:

$AC = a$, the major axis of the ellipse
$BC = b$, the minor axis of the ellipse
AT is the tangent to the ellipse at A
CT cuts the ellipse at M
$AM = s$ is the length of the arc AM
$AT = t$
$CP = x$

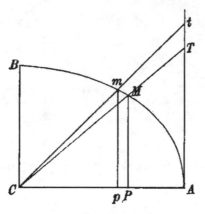

Figure 1.

Euler plans to use differentials on this diagram, so he means the distance tT and the arc mM to be very small. Also, we are to assume that mp and MP are perpendicular to the axis CA.

Note that Euler uses the variable t twice here, once as a point and once as a length. This was a common practice in the eighteenth century and it often gets confusing. Just as Euler sometimes uses the same name for two different objects, he also uses two different names for the same thing. We will see him call the arc BMA sometimes, and call it AMB other times, and he seems to switch randomly between $bb + tt$ and $b^2 + t^2$. We will try to stick with the way Euler wrote it and let the reader puzzle about whether there is a pattern.

We are talking about an ellipse that today we would describe with the equation

$$\frac{x^2}{a^2} + \frac{y^2}{b^2} = 1.$$

Euler gives an equivalent form, $PM = b\sqrt{a^2 - x^2}/a$, and notes that, by similar triangles, $tx = b\sqrt{a^2 - x^2}$ or, equivalently, $x = ab/\sqrt{bb + tt}$.

With a bit of work, we can take $y = PM$ and knowing that the arc length differential is given by $ds = \sqrt{1 + (y')^2}$ we find that

$$ds = \frac{-dx\sqrt{a^4 - (a^2 - b^2)x^2}}{a\sqrt{a^2 - x^2}}.$$

We could try to integrate this between 0 and a to find the length of the arc BMA, but Euler and others have learned from experience that this doesn't work very well. Instead, Euler replaces x with t. This gives

$$ds = \frac{b\,dt\sqrt{b^4 + a^2tt}}{(bb + tt)^{3/2}}.$$

Now the arc length BMA is the integral of this from 0 to ∞.

Before he gets down to the integral, Euler wants to make one more substitution. If you think about it, the ratio of the axes of an ellipse, a/b tells us how much the ellipse is like a circle. If the ratio is close to 1, then the ellipse is more circular. Euler wants, instead, a measure of how *different* the ellipse is from a circle. He defines a measure n by the equation $a^2 = (n + 1)b^2$. Here, when n is close to zero, then a is close to b and the ellipse is not much different from a circle. If we use this to replace a with n, we get

$$ds = \frac{b^2 \, dt \, \sqrt{(b^2 + t^2) + nt^2}}{(b^2 + t^2)^{3/2}}.$$

This doesn't look like progress, but Euler has a surprise, one that Newton had used over 50 years earlier. Remember the binomial theorem:

$$(x + y)^m = x^m + \binom{m}{1} x^{m-1} y + \binom{m}{2} x^{m-2} y^2 + \binom{m}{3} x^{m-3} y^3 + \cdots$$

$$= x^m + \frac{m}{1} x^{m-1} y + \frac{m(m-1)}{1 \cdot 2} x^{m-2} y^2 + \frac{m(m-1)(m-2)}{1 \cdot 2 \cdot 3} x^{m-3} y^3 + \cdots$$

We usually use this for m a positive integer, and, in that case, the numerators in the second formula eventually are zero and we get a finite series. Newton showed that the theorem is still true for fractional values of m, but the result is an infinite series. Euler takes $m = \frac{1}{2}$ and denotes the coefficients by

$$A = \frac{1}{2}$$

$$B = \frac{\frac{1}{2}\left(\frac{1}{2} - 1\right)}{1 \cdot 2} = \frac{-1}{2} \cdot \frac{1}{4}$$

$$C = \frac{\frac{1}{2}\left(\frac{1}{2} - 1\right)\left(\frac{1}{2} - 2\right)}{1 \cdot 2 \cdot 3} = \frac{1 \cdot 1 \cdot 3}{2 \cdot 4 \cdot 6}, \text{ etc.}$$

Now he applies the binomial theorem to the radical in the numerator of ds to get

$$(b^2 + t^2)^{1/2} + \frac{Ant^2}{(b^2 + t^2)^{1/2}} + \frac{Bn^2t^4}{(b^2 + t^2)^{3/2}} + \frac{Cn^3t^6}{(b^2 + t^2)^{5/2}} + \text{ etc.}$$

so that

$$ds = \frac{b^2 dt}{b^2 + t^2} + \frac{Ab^2nt^2 \, dt}{(b^2 + t^2)^2} + \frac{Bb^2n^2t^4 \, dt}{(b^2 + t^2)^3} + \frac{Cb^2n^3t^6 \, dt}{(b^2 + t^2)^4} + \text{ etc.}$$

So, the length of the arc AMB will be the integral of this series from 0 to ∞. Notice how nicely Euler's trick got rid of the radical in the denominator, and also how introducing the term n helps the series converge rapidly for small values of n.

The first term of this series integrates as an arctangent to give $b\frac{\pi}{2}$. The rest of the terms reduce, as if by magic, to the first term using integration by parts. The second term, for example, (ignoring A and n to make it a little easier to type) gives

$$\int \frac{b^2t^2 \, dt}{(b^2 + t^2)^2} = \frac{1}{2} \int \frac{bb \, dt}{bb + tt} - \frac{1}{2} \frac{b^2t}{bb + tt}$$

where in the integration by parts, we took

$$dv = \frac{2t \, dt}{(b^2 + t^2)^2} \quad \text{so that} \quad v = \frac{-1}{b^2 + t^2}.$$

Similarly, the third term reduces to the second, and the fourth to the third, and we get, after a bit of calculation,

$$\int \frac{b^2 t^4\, dt}{(b^2 + t^2)^3} = \frac{1\cdot 3}{2\cdot 4}\int \frac{b^2\, dt}{bb + tt} - \frac{1\cdot 3}{2\cdot 4}\frac{b^2 t}{bb + tt} - \frac{1}{4}\frac{b^2 t^3}{(bb + tt)^2}$$

and

$$\int \frac{b^2 t^6\, dt}{(b^2 + t^2)^4} = \frac{1\cdot 3\cdot 5}{2\cdot 4\cdot 6}\int \frac{b^2\, dt}{bb + tt} - \frac{1\cdot 3\cdot 5}{2\cdot 4\cdot 6}\frac{b^2 t}{bb + tt} - \frac{1\cdot 5}{4\cdot 6}\frac{b^2 t^3}{(bb + tt)^2} - \frac{1}{6}\frac{b^2 t^5}{(bb + tt)^3}.$$

Euler is a genius at such calculations, and he tells us that from this, "the law for the integrals of the remaining terms is apparent enough."

The information Euler needs to find the length of the arc AMB, is hidden in these series. He points out that when $t = 0$ or $t = \infty$, the "algebraic" terms, that is, the terms outside the integrand, are all zero, so we only have to worry about the integrals themselves. With those swept away, the pattern for the reduction of the integral becomes clear:

$$\int \frac{b^2 t^{2m}\, dt}{(b^2 + t^2)^{m+1}} = \frac{1\cdot 3\cdot 5\cdots (2m-1)}{2\cdot 4\cdot 6\cdots 2m}\frac{\pi b}{2}.$$

We are almost done. Euler takes $e = \frac{\pi b}{2}$. This looks odd to us, but Euler had not yet adopted the convention that the symbol e always denotes the base for the natural logarithms. Substitute these values back in the integral of ds, and putting the parameter n and the coefficients A, B, C, etc., back into the equation, we get

$$AMB = e\left(1 + \frac{1}{2}An + \frac{1\cdot 3}{2\cdot 4}Bn^2 + \frac{1\cdot 3\cdot 5}{2\cdot 4\cdot 6}Cn^3 + \frac{1\cdot 3\cdot 5\cdot 7}{2\cdot 4\cdot 6\cdot 8}Dn^4 + \text{etc.}\right).$$

If we also substitute for A, B, C, D, we get

$$AMB = e\left(1 + \frac{1\cdot 1\cdot n}{2\cdot 2} - \frac{1\cdot 3\cdot n^2}{2\cdot 2\cdot 4\cdot 4} + \frac{1\cdot 1\cdot 3\cdot 3\cdot 5\cdot n^3}{2\cdot 2\cdot 4\cdot 4\cdot 6\cdot 6} - \frac{1\cdot 1\cdot 3\cdot 3\cdot 5\cdot 5\cdot 7\cdot n^4}{2\cdot 2\cdot 4\cdot 4\cdot 6\cdot 6\cdot 8\cdot 8} + \text{etc.}\right).$$

This is Euler's answer, a rather intimidating series. We might want to replace the symbol e with its value and write it in terms of π as

$$AMB = \frac{\pi b}{2}\left(1 + \frac{1\cdot 1\cdot n}{2\cdot 2} - \frac{1\cdot 3\cdot n^2}{2\cdot 2\cdot 4\cdot 4} + \frac{1\cdot 1\cdot 3\cdot 3\cdot 5\cdot n^3}{2\cdot 2\cdot 4\cdot 4\cdot 6\cdot 6} - \frac{1\cdot 1\cdot 3\cdot 3\cdot 5\cdot 5\cdot 7\cdot n^4}{2\cdot 2\cdot 4\cdot 4\cdot 6\cdot 6\cdot 8\cdot 8} + \text{etc.}\right).$$

We can check that, when $n = 0$ and $a = b = 1$, we get the answer we expect, $\pi/2$, and that as a increases, n also increases, as does the value of the series, so the answer at least makes sense, even if it isn't as simple as we might have hoped.

Over the last 270 years, we have learned a lot more about arc lengths on ellipses. Euler himself added a good deal more to the subject, including the so-called addition formula for elliptic integrals. These arc lengths are the foundation of deep and rich studies of elliptic integrals, elliptic curves and elliptic functions, with applications across a vast spectrum of mathematics, from mechanics to Wiles' solution of Fermat's Last Theorem. It all has roots in this paper, and the curious fact that arc length for ellipses is so much more complicated than it is for circles.

References

[E28] Euler, Leonhard, Specimen de constructione aequationum differentialium sine indeterminatarum separatione, *Commentarii academiae scientiarum Petropolitanae* 6 (1732/33) 1738, pp. 168–174, reprinted in *Opera Omnia* Series I vol. 20 pp. 1–7. Available online at `EulerArchive.org`.

28

Mixed Partial Derivatives

(May 2004)

One of the first things we learn in Calculus III, multivariable calculus, is that mixed partial derivatives are equal. That is, for most familiar functions of two variables, say $f(x, y)$, it doesn't matter whether you take partial derivatives first with respect to x, then with respect to y, or if you do it the other way around. In symbols, this says

$$\frac{\partial^2 f}{\partial x \partial y} = \frac{\partial^2 f}{\partial y \partial x}, \quad \text{or} \quad f_{xy} = f_{yx}.$$

Almost the next thing we learn is that there are a few conditions of continuity that our function $f(x, y)$ must satisfy to assure this equality. We learn a number of special counterexamples, which, I for one, remembered for the test, but then forgot until I had to teach Calculus III myself.

Two hundred and seventy years ago, this fact about partial derivatives was unknown. The very idea of functions was new. People used equations of two variables to describe curves, and of three variables to describe surfaces, but they hadn't made the transition from equations to functions. Surfaces were additionally difficult to deal with because three-dimensional coordinate systems were new and people were not yet comfortable with them.

Somehow, in the midst of this confusion, Euler was able to discover the fact that mixed partial derivatives are equal. Since Euler did not know any of the functions that could have served as counterexamples, we should not begrudge it that he did not also discover the continuity conditions.

We will explore Euler's discovery by looking at three questions. First, without the tools of surfaces, functions and three-dimensional coordinate systems, how could Euler make this discovery? Second, exactly what did he discover, anyway? And finally, what kind of proof or evidence did he offer to make people believe it was true?

The answer to the first question is a little surprising. Euler wasn't thinking about surfaces, functions or three-dimensional coordinate systems when he wrote this paper in 1734. The paper whose title in English is "On an infinity of curves of a given kind, or a method of finding equations for an infinity of curves of a given kind" is number 44

[E44] on Eneström's index of Euler's works. The title is as awkward in its original Latin as it is in English. As the title suggests, the paper is about families of curves. The title does not suggest that Euler means to study the differential equations satisfied by a given family of curves.

Euler begins his paper with a discussion of what we would call a *parameter*, but Euler calls a *modulus*. It is a relatively rare example of a term that Euler used that was not adopted by the rest of the mathematical community. He uses an example, $y^2 = ax$, which he interprets as describing infinitely many parabolic curves, one for each value of a, and all with the same axis and vertex. He intends to examine how the curves change as the value of a changes. So we see that Euler was thinking about curves, parameters and two-dimensional coordinate systems, not surfaces, functions and three-dimensional coordinate systems.

What was he thinking about? At the time, people didn't say they were "solving" a differential equation. Instead, they said they were "integrating" or "constructing" it, so it was natural for a study of differential equations to begin with integration. Euler asks us to consider $z = \int P\,dx$, where P involves a, z and x. Then $dz = P\,dx$, in which expression a is to be considered as a constant. Euler works hard to explain that if a is considered a variable, then this last expression could be differentiated with respect to a, and also that if $dz = P\,dx$ is integrated, then the resulting expression might involve a function of a. All this leads to a conclusion that seems paradoxical when we first see it, that if a is a constant, then

$$dz = P\,dx$$

but if a is considered as a variable, then

$$dz = P\,dx + Q\,da.$$

Euler moves on to state a theorem: "*If a quantity A composed of two variables t and u is differentiated first holding t constant, and then that differential is differentiated holding u constant and letting t be a variable, then the same result will occur if the order is reversed and A is first differentiated holding u constant and then that differential is differentiated holding t constant and letting u be a variable.*"

We recognize this as claiming that mixed partial derivatives are equal, regardless of the order of the differentiation. Today, this is the second thing we learn about partial derivatives when we encounter them in calculus class. There, we also learn about some continuity conditions that Euler does not yet know about.

The awkward wording and the lack of notation are dictated by Euler's times. He writes about differentials and not derivatives, so the very idea of partial derivatives and second order partial derivatives is more difficult to discuss. Our modern notation for partial derivatives has evolved over many years specifically so that it is easy to use it to write facts like this. Compare

$$\frac{\partial^2 f}{\partial x \partial y} = \frac{\partial^2 f}{\partial y \partial x} \quad \text{or} \quad f_{xy} = f_{yx}$$

to the tools Euler had to make this same statement.

Our third question was about Euler's proof. For this, we can translate Euler himself:

Suppose A is a function of t and u. From A we get B if, in place of t in A we substitute $t + dt$; and we get C if in place of u in A we substitute $u + du$. If we simultaneously substitute $t + dt$ and $u + du$, we change A into D. From a different point of view, we could get D by substituting $u + du$ for u in B or by substituting $t + dt$ for t in C. This said, if the differential of A is taken, holding t constant, it will produce $C - A$. If in $C - A$ we put $t + dt$ in place of t, it will produce $D - B$, the differential of which will be

$$D - B - C + A.$$

Now, doing things in the other order, if $t + dt$ is put into A in place of t, then B is produced, and then the differential of A, taking t to be the variable, will be $B - A$. Putting $u + du$ in place of u in this differential will give

$$D - B - C + A,$$

which is equal to the differential found in the previous operations.   Q.E.D.

All of this fills only three pages of Euler's 20-page paper. The rest of the paper is concerned with some now-forgotten questions of how an integral like $z = \int P \, dx$ depends on a, if P is a function of x and also involves a parameter a. He takes the differential, $dz = P \, dx + Q \, da$, and Euler wants to understand what Q is. Euler takes more differentials, $dP = A \, dx + B \, da$ and $dQ = C \, dx + D \, da$. Because of Euler's result about mixed partial derivatives, he knows that $C = B$, so $dQ = B \, dx + D \, da$, and finally, $Q = \int B \, dx$. This part of the paper goes on and on, as Euler considers different forms for P and adds more and more complications.

It is a clear case for which the tool is more important than the problem.

There are two other minor features of this paper. The first involves e, the base for natural logarithms. Euler pioneered the use of standard symbols for such constants. He was also responsible for our universal use of the symbol π. Euler's first "official" use of e was in his two volume *Mechanica*, published in 1736. This paper was *written* in 1734, but not published until 1740. That delay in publication cost it first place.

Second, some sharp-eyed readers may have noticed that this paper occupies pages 174–189 and 180–183 of its volume of the *Commentarii*. It's not our typographical error. A typesetter in 1740 got confused as he set the page numbers for this volume, and after page 189 he put another page 180.

References

[E44] Euler, Leonhard, De infinitis curvis eiusdem generis seu methodus inveniendi aequationes pro infinitis curvis eiusdem generis, *Commentarii academiae scientiarum Petropolitanae* 7 (1734/5) 1740, pp. 174–189, 180–183, reprinted in *Opera Omnia* Series I vol. 22 pp. 36–56. Available online at EulerArchive.org.

29

Goldbach's Series

(February 2005)

Christian Goldbach (1690–1764) is a fairly well known, but rather minor figure in the history of mathematics. Pictures of Goldbach seem very rare, if indeed any survive. His name, if not his face, is widely known because it is attached to his conjecture that every even number larger than two is the sum of two prime numbers. Of the great unsolved problems in mathematics, the Goldbach Conjecture is probably the easiest to explain to a non-mathematician. Try explaining the Riemann hypothesis, for example, to a fourth grader or an English major.

Goldbach was also a kind of mentor to Leonhard Euler. For over 25 years they exchanged letters, 196 of which survive. These letters give us a window into Euler's scientific and personal life. In Goldbach's very first letter to Euler, dated December 1, 1729, Goldbach got Euler interested in number theory. Goldbach added a note at the end of the letter: "P. S. Have you noticed the observation of Fermat that all numbers of the form $2^{2^{x-1}} + 1$, that is 3, 5, 17, etc., are prime numbers, but he did not dare to claim he could demonstrate it, nor, as far as I know, has anyone else been able to prove it." This was mentioned in this column in November 2003.

Euler was not the only one to benefit from Goldbach's attentions. Earlier in 1729, Goldbach posed two series to Euler's friend Daniel Bernoulli. Those series were:

the sum of the reciprocals of the numbers one less than a power of two, that is:

$$1 + \frac{1}{3} + \frac{1}{7} + \frac{1}{15} + \cdots + \frac{1}{2^n - 1} + \text{etc., and}$$

the sum of the reciprocals of the numbers one less than powers, as in Theorem 1 below, that is:

$$\frac{1}{3} + \frac{1}{7} + \frac{1}{8} + \frac{1}{15} + \frac{1}{24} + \frac{1}{26} + \text{etc.}$$

Euler mentions both of these series in E25, written in 1732 and published in 1738. Sometime after Euler wrote that paper and before 1737 when he wrote E72, Euler learned that Goldbach had found a way to sum the second of these series. In this column, we will

167

look at some of the results in that paper, E72, *Variae observationes circa series infinitas*, or "Several observations about infinite series." We will see what Goldbach did, and see how much more Euler could do with the same idea. We will save some of the results for a later column.[1]

E72 is Euler's first paper that closely follows the modern Theorem-Proof format. There are no definitions in the paper, or it would probably follow the Definition-Theorem-Proof format. After an introductory paragraph in which Euler tells part of the story of the problem, Euler gives us a theorem and a proof:

Theorem 1. *This infinite series, continued to infinity,*

$$\frac{1}{3} + \frac{1}{7} + \frac{1}{8} + \frac{1}{15} + \frac{1}{24} + \frac{1}{26} + \frac{1}{31} + \frac{1}{35} + etc.$$

the denominators of which are all numbers which are one less than powers of degree two or higher of whole numbers, that is, terms which can be expressed with the formula $1/(m^n - 1)$, where m and n are integers greater than one, then the sum of this series is 1.

Euler's proof begins with an 18th century step that treats infinity as a number. Such steps became unpopular among rigorous mathematicians about a hundred years later. He takes x to be the sum of the harmonic series

$$x = 1 + \frac{1}{2} + \frac{1}{3} + \frac{1}{4} + \frac{1}{5} + \frac{1}{6} + \frac{1}{7} + \frac{1}{8} + \frac{1}{9} + etc.$$

Today, Euler would have learned in calculus that we are not allowed to treat x as an infinite number like this, and he would not have discovered this remarkable proof.

Next, Euler subtracts the geometric series

$$1 = \frac{1}{2} + \frac{1}{4} + \frac{1}{8} + \frac{1}{16} + \frac{1}{32} + etc.$$

leaving

$$x - 1 = 1 + \frac{1}{3} + \frac{1}{5} + \frac{1}{6} + \frac{1}{7} + \frac{1}{9} + \frac{1}{10} + etc.$$

Subtract another geometric series

$$\frac{1}{2} = \frac{1}{3} + \frac{1}{9} + \frac{1}{27} + \frac{1}{81} + \frac{1}{243} + etc.$$

leaving

$$x - 1 - \frac{1}{2} = 1 + \frac{1}{5} + \frac{1}{6} + \frac{1}{7} + \frac{1}{10} + \frac{1}{11} + etc.$$

and another geometric series

$$\frac{1}{4} = \frac{1}{5} + \frac{1}{25} + \frac{1}{125} + etc.$$

leaving

$$x - 1 - \frac{1}{2} - \frac{1}{4} = 1 + \frac{1}{6} + \frac{1}{7} + \frac{1}{10} + etc.$$

Note that Euler had to skip subtracting the geometric series

$$\frac{1}{3} = \frac{1}{4} + \frac{1}{16} + \frac{1}{64} + \frac{1}{256} + \text{etc.}$$

because the series of powers of $\frac{1}{4}$ on the right is already a subseries of the series of powers of $\frac{1}{2}$, so those terms have already been subtracted. This happens because 3 is one less than a power, 4. It happens again every time we reach a term one less than a power. He will have to skip 7, because that is one less than the cube 8, 8 because it is one less than the square 9, 15 because it is one less than the square 16, etc.

Continuing in this way, we see that all of the terms on the right except the term 1 can be eliminated, leaving

$$x - 1 - \frac{1}{2} - \frac{1}{4} - \frac{1}{5} - \frac{1}{6} - \frac{1}{9} - \text{etc.} = 1$$

so

$$x - 1 = 1 + \frac{1}{2} + \frac{1}{4} + \frac{1}{5} + \frac{1}{6} + \frac{1}{9} + \frac{1}{10} + \text{etc.}$$

Now it gets just a little bit tricky. Since x is the sum of the harmonic series, Euler believes that the 1 on the left must equal the terms of the harmonic series that are missing on the right. Those missing terms are exactly the ones with denominators one less than powers, so Euler concludes that

$$1 = \frac{1}{3} + \frac{1}{7} + \frac{1}{8} + \frac{1}{15} + \frac{1}{24} + \frac{1}{26} + \text{etc.}$$

where the terms on the right have denominators one less than powers.   Q.E.D.

This is a beautiful proof, and a true result. Unfortunately, the beautiful proof is of the form "infinity minus infinity equals 1," and so is not valid. Bruce Burdick, of Roger Williams University, has discovered, but not yet published, a direct proof, using the principle of inclusion and exclusion, that meets modern standards of rigor, but it somehow lacks the spirit of Goldbach's proof, as presented by Euler. Also, Pelegrí Viader of Barcelona has discovered a proof that uses non-standard analysis. He says that his paper will appear soon.[2]

This seems to be as far as Goldbach took this problem, and it's a little hard to know how much of this proof is Euler's and how much is Goldbach's. Euler has more ideas. He gives us

Theorem 2. $\frac{1}{3} + \frac{1}{7} + \frac{1}{15} + \frac{1}{31} + \frac{1}{35} + \frac{1}{63} + \text{etc.} = l\,2.$

Here the denominators are one less than the powers of even numbers, and Euler, as we have seen before, writes $l\,2$ to denote what we would write $\ln 2$.

In Theorem 3, he tells us that

$$\frac{\pi}{4} = 1 - \frac{1}{8} - \frac{1}{24} + \frac{1}{28} - \frac{1}{48} - \frac{1}{80} - \frac{1}{120} - \frac{1}{124} - \frac{1}{168} - \frac{1}{224} + \frac{1}{244} - \frac{1}{288} - \text{etc.}$$

---

[2]Since this column was written in February 2005, Viader's paper [BPV] appeared in *The American Mathematical Monthly*. Burdick has yet to publish his paper.

where the denominators are "evenly even" numbers, that is numbers that are divisible by four, that are one more or one less than the powers of odd numbers, and where the denominators that are one more than powers have a + sign, while the denominators that are one less have a − sign.

Theorems 4, 5 and 6 are even more exotic series.

We are only about half way through this paper, and there are remarkable things to come. Euler will adapt these same methods to develop theorems relating infinite sums to infinite products, prove an important result about what we now call the Riemann zeta function, and give an entirely new proof of the result, first given by Euclid, that there are infinitely many prime numbers. We will examine these results in a later column.

References

[D]   Dunham, William, *Euler: The Master of Us All*, Dolciani Mathematical Expositions vol. 22, Mathematical Association of America, Washington, DC, 1999.

[BPV] Bibiloni, Lluís, Jaume Paradís and Pelegrí Viader, On a series of Goldbach and Euler, *American Mathematical Monthly*, 113, 3 (March 2006) pp. 206–220.

30

Bernoulli Numbers

(September 2005)

As we learned in last month's column, in the 1760s Euler wrote only two articles on series. There was E326, a paper on the so-called central trinomial coefficients and the subject of that column. In researching that column, I also looked at the other paper, E393, *De summis serierum numeros Bernoullianos involventium*, or "On sums of series involving Bernoulli numbers." While the paper itself didn't turn out to be all that interesting, the path that leads to the paper is fascinating. It is the story of the Bernoulli numbers.

Bernoulli numbers are a sequence of rational numbers that arise in a dazzling variety of applications in analysis, numerical analysis and number theory. When Charles Babbage designed the Analytical Engine in the 19th century, one of the most important tasks he hoped the Engine would perform was the calculation of Bernoulli numbers.

The first few Bernoulli numbers [K] are

$$B_0 = 1$$
$$B_1 = \frac{-1}{2}$$
$$B_2 = \frac{1}{6}$$
$$B_3 = 0$$
$$B_4 = \frac{-1}{30}$$
$$B_5 = 0$$
$$B_6 = \frac{1}{42}$$

After B_1 all Bernoulli numbers with odd index are zero, and the nonzero ones alternate in sign. They first appeared in 1713 in Jakob Bernoulli's pioneering work on probability, *Ars Conjectandi*. Jakob Bernoulli (1654–1705) was the older brother of Johann Bernoulli (1667–1748), who was, in turn, Euler's teacher and mentor at the University of Basel.

Sometimes people simply omit the Bernoulli numbers with odd index from the list,

JACOBI BERNOULLI,
Profeſſ. Baſil. & utriuſque Societ. Reg. Scientiar.
Gall. & Pruſſ. Sodal.
MATHEMATICI CELEBERRIMI,

ARS CONJECTANDI,

OPUS POSTHUMUM.

Accedit

TRACTATUS

DE SERIEBUS INFINITIS,-

Et EPISTOLA Gallicè ſcripta

DE LUDO PILÆ
RETICULARIS.

FESTINA LENTE

BASILEÆ,
Impenſis THURNISIORUM,·Fratrum.
cIɔ Iɔcc XIII.

and write B_k^* where we write B_{2k}. They, of course, must then make certain modifications to their formulas, and, in general, their formulas are a bit simpler.

Bernoulli was studying sums of powers of consecutive integers, like sums of squares,

$$1 + 4 + 9 + 16 + 25 = 55$$

or sums of cubes

$$1 + 8 + 27 + 64 + 125 + 216 + 343 = 784$$

In modern notation (Bernoulli did not use subscripts, nor did he use Σ for summations or ! for factorials) Bernoulli found that

$$\sum_{k=1}^{n-1} k^p = \sum_{k=0}^{p} \frac{B_k}{k!} \frac{p!}{(p+1-k)!} n^{p+1-k}.$$

If n is large and p is small, that means that the left-hand side is a sum of a relative large number of relatively small powers, and if we know the necessary Bernoulli numbers then the sum on the right is simpler to evaluate than the sum on the left. Bernoulli himself is said [GS] to have used this formula to find the sum of the tenth powers of numbers 1 to 1000 in less than eight minutes. The answer is a 32-digit number.

Bernoulli numbers arise in Taylor series in the expansion

$$\frac{x}{e^x - 1} = \sum_{k=0}^{\infty} B_k \frac{x^k}{k!}.$$

Bernoulli numbers are also involved in the expansions of several other functions, including $\tan x$, $\frac{x}{\sin x}$, $\log\left(\frac{\sin x}{x}\right)$ and others.

Euler encountered Bernoulli numbers in his great solution to the Basel problem when he showed that

$$\sum_{k=1}^{\infty} \frac{1}{k^2} = 1 + \frac{1}{4} + \frac{1}{9} + \frac{1}{16} + \frac{1}{25} + \text{etc.} = \frac{\pi^2}{6},$$

though he did not recognize them at the time. In the same paper, Euler also evaluated

$$\sum_{k=1}^{\infty} \frac{1}{k^n}$$

for the first several even values of n. Only later would he realize that these other sums involved the Bernoulli numbers. In fact, if n is even, then

$$\sum_{k=1}^{\infty} \frac{1}{k^n} = \frac{2^n |B_n| \pi^n}{2(n!)}.$$

Euler also failed to recognize the Bernoulli numbers in 1732 when he first did his work on the Euler-Maclaurin formula. Maclaurin also missed them when he discovered the formula independently in 1742. Again in modern form, the result says that for sufficiently smooth functions f, a series based on f and an integral of f are related by

$$\sum_{k=1}^{n} f(k) = \int_{1}^{n} f(x)\, dx + \frac{f(1) + f(n)}{2} + \sum_{k=1}^{p} \frac{B_{2k}}{(2k)!}\left(f^{2k-1}(n) - f^{2k-1}(1)\right) + R_n(f, p)$$

where $R_n(f, p)$ is a remainder term that usually disappears rapidly as p increases. [GS] The formula can be used either to estimate the series on the left knowing the integral on the right, or conversely, to estimate the integral by evaluating the series. Euler used the series on the left-hand side of this formula in 1732 to estimate the values of infinite series and to find $\sum_{k=1}^{\infty} 1/k^2$ to six decimal places. This is also how he found the first properties of γ, the so-called Euler constant. Maclaurin used the other side of the formula to estimate the values of integrals from series.

The Bernoulli numbers are related to Euler's constant γ by

$$\gamma = \frac{1}{2} + \sum_{k=1}^{\infty} \frac{B_{2k}}{2k}.$$

There is also an astonishing result due to Eduard Kummer (1810–1893) [GS] relating Bernoulli numbers to Fermat's Last Theorem. Kummer noticed that a prime number p is special if it does not divide the numerators of any of the Bernoulli numbers $B_2, B_4, B_6, \ldots, B_{p-3}$. Such primes are now called *regular*. The first prime that is *not* regular is 37. Kummer showed that if p is a regular prime, then Fermat's Last Theorem, that $x^p + y^p = z^p$ has no nontrivial integer solutions, is true for p.

In 1921, Eric Temple Bell, author of the well-known popular mathematics history book *Men of Mathematics*, proved [B]:

Theorem.   *If p is an odd prime which does not divide $4^r - 1$, then the numerator of B_{2pr} is divisible by p.*

There must be thousands of such results, and Bernoulli numbers continue to be studied today. JSTOR reports over 150 "hits" on the key words "Bernoulli numbers" since 1990.

Let us turn now to 1755, when Euler published his *Institutiones calculi differentialis* [E212]. At that time, only a few of the results above were known, and their links to Bernoulli numbers were apparently not yet recognized. The results that were known seem to be:

1. Bernoulli's own results on summing powers of integers. Bernoulli showed how this involved Bernoulli numbers, hence the name.

2. The Euler-Maclaurin summation formula.

3. Taylor series for various functions.

4. Euler's evaluation of $\zeta(2n)$.

Then, through all the trees, Euler sees the forest. It must have been a wonderful feeling to see how so many different aspects of mathematics are linked through these mysterious Bernoulli numbers.

Euler devotes almost all of chapters 5 and 6 of Part 2 of his *Calculi differentialis* to results related to Bernoulli numbers, and on page 420 (page 321 of the *Opera Omnia* edition) he attributes them to Jakob Bernoulli and calls them *Bernoulli numbers*. Unfortunately, only Part 1 of the *Calculi differentialis* has been translated into English, so readers who want to enjoy it in Euler's words must either brave the Latin or find a copy of the rare 1790 German translation.

Euler begins his chapter 5, "Investigation of the sums of series from their general term" with a quick treatment of Bernoulli's results on summing sequences of powers. Then he repeats his own results from the 1730s [E25] on the Euler-Maclaurin formula and gives the recursive relation on the coefficients in that formula. Euler doesn't mention Maclaurin, so he is probably unaware of his work on the subject.

Then he shows how those coefficients arise from the Taylor series expansions of

$$\frac{x}{1 - e^{-x}} \quad \text{and} \quad \frac{1}{2} \cot\left(\frac{1}{2}x\right).$$

Eventually, after quite a bit of work, he lists the Bernoulli numbers, naming them after Bernoulli in the process, and shows how they are related to the coefficients in the Euler-Maclaurin formula.

This done, he extends occurrence of Bernoulli numbers in the expansion of $\frac{1}{2}\cot\left(\frac{1}{2}x\right)$ to the more general form $\frac{\pi}{n}\cot\left(\frac{m\pi}{n}\right)$ and uses that to relate Bernoulli numbers to the values of $\zeta(2n)$. To end the theoretical parts of his exposition, he gives some of the properties of the Bernoulli polynomials and notes that Bernoulli numbers grow faster than any geometric series.

Euler spends the rest of these two chapters doing applications of Bernoulli numbers, including calculating the Euler-Mascheroni constant, γ, to 15 decimal places.

All this is rather unexpected in a textbook on differential calculus.

With this, Euler did not write again on Bernoulli numbers until 1768. In fact, in the intervening 13 years, he wrote only four papers on series, one on central trinomial coefficients that was the subject of another column, one paper on approximating pi, one on trig functions, and one on continued fractions.

We have already said that the 1768 paper, E393, "didn't turn out to be all that interesting," but it might be worth summarizing its results. Euler opens E393 with a list of Bernoulli numbers and a list of the coefficients that arise in $\zeta(2n)$, and shows how the two lists are related. Then he gives his recursive relation on the zeta coefficients.

Then he leaps to the Euler-Maclaurin formula. Up to this point, most of the essay is just a new version of what he had presented in the *Calculi differentialis*. From here, though, he gives a different way to show the relation between the Bernoulli numbers and the expansion of $\frac{1}{2}-\frac{1}{2}\cot\left(\frac{1}{2}x\right)$. Then he uses this same technique to give new relations between the Bernoulli numbers and a variety of other functions and numbers, including

$$\frac{x}{2}\frac{e^y+e^{-y}}{e^y-e^{-y}}-\frac{1}{2}\quad\text{and}\quad\frac{1}{e^2-1}.$$

Finally, he gives the values of integrals like

$$\int_0^1\frac{(\ln x)^n}{(1-x)}\,dx,$$

for n odd, in terms of the $(n+1)$st Bernoulli number. None of this would have been appropriate to include in the *Calculi differentialis*.

Simon Singh [S] quotes Andrew Wiles as describing the process of mathematical discovery with the colorful words "You enter the first room of the mansion and it's completely dark. You stumble around bumping into the furniture but gradually you learn where each piece of furniture is. Finally, after six months or so, you find the light switch, you turn it on, and suddenly it's all illuminated." It must have been something like this for Euler, when he saw how the "furniture" was arranged around the Bernoulli numbers.

In Euler's time, though, light switches hadn't yet been invented.

References

[B]   Bell, E. T., Note on the Prime Divisors of the Numerators of Bernoulli's Numbers, *The American Mathematical Monthly*, Vol. 28, 6/7 (Jun.–Jul. 1921), pp. 258–259.

[E25]   Euler, Leonhard, Methodus generalis summandi progressiones, *Commentarii academiae scientiarum Petropolitanae* 6 (1732/3) 1738, pp. 68–97. Reprinted in *Opera Omnia* Series I vol. 14, pp. 42–72. Available online at EulerArchive.org.

[E212] ——, *Institutiones calculi differentialis*, St. Petersburg, 1755, reprinted in *Opera Omnia* Series I vol. 10. English translation of Part I by John Blanton, Springer, NY, 2000. Available online at EulerArchive.org.

[E393] ——, De summis serierum numeros Bernoullianos involventium, *Novi commentarii academiae scientiarum Petropolitanae* 14 (1769): I, 1770, pp. 129–167, reprinted in *Opera Omnia* Series I vol. 15 pp. 91–130. Available online at EulerArchive.org.

[GS] Gourdon, Xavier and Pascal Sebah, Numbers, constants and computation, online at numbers.computation.free.fr/Constants/constants.html, link to Constants, Miscellaneous, Bernoulli numbers, consulted July 25, 2005.

[H] Havil, Julian, *Gamma: Exploring Euler's Constant*, Princeton University Press, 2003.

[K] Kellner, Bernd C., The Bernoulli Number Page online at Bernoulli.org. Consulted July 25, 2005.

[S] Singh, Simon, Who is Andrew Wiles?, SimonSingh.net, on line at simonsingh.com/Andrew_Wiles.html. Consulted August 1, 2005.

31

Divergent Series

(June 2006)

Today we are fairly comfortable with the idea that some series just don't add up. For example, the series

$$1 - 1 + 1 - 1 + 1 - 1 + \text{etc.}$$

has nicely bounded partial sums, but it fails to converge, in the modern meaning of the word. It took mathematicians centuries to resolve the paradoxes of diverging series, and this month's column is about an episode while we were still confused.

In the 1700s, though, many mathematicians were more optimistic, or perhaps more naïve, about the limitations of mathematics, thinking that with enough brilliance and enough work they could solve any differential equation and sum any series. Daniel Bernoulli, for example, thought that the series above ought to have value $\frac{1}{2}$, not for the usual reason that involves geometric series and $\frac{1}{1-x} = 1 + x + x^2 + x^3 + \cdots$, but because of a probabilistic argument. He thought that since half of the partial sums of the series are $+1$ and half of them are zero, the correct value of the series would be the expected value $\frac{1}{2} \cdot 1 + \frac{1}{2} \cdot 0 = \frac{1}{2}$.

By a similar argument, Bernoulli concluded that $1 + 0 - 1 + 1 + 0 - 1 + 1 + 0 - 1 +$ etc. would have value $\frac{2}{3}$, and that, by a judicious interpolation of zeroes into the series, it could be argued to have any value between 0 and 1.

It is sometimes frustrating to us that the writers of the time usually did not distinguish between a series and a sequence. The words "progression" and "series" took both meanings. At the same time, though, they made a now-obsolete distinction between the *value* of a series and the *sum* of the same series. Bernoulli and his contemporaries would give the series $1 - 1 + 1 - 1 + 1 - 1 + \text{etc.}$ the *value* $\frac{1}{2}$, since they could make reasonable calculations and analyses that supported this value, but they would be reluctant to call that value a *sum*. They seemed to think that sums required convergence of some sort.

Euler entered the conversation with a paper *De seriebus divergentibus* (On divergent series) [E247], written in 1746, but not read to the Academy until 1754, nor published until 1760. Euler wrote this paper about two years after he finished his great precalculus textbook the *Introductio in analysin infinitorum*, in which he devotes a good deal of time to issues of series. The *Introductio* does not deal with divergent series directly, but they are often near to his thoughts.

Euler's intent in E247 is to give a value to a series he calls the "hypergeometric series of Wallis:"

$$1 - 1 + 2 - 6 + 24 - 120 + 720 - 5040 + \text{etc.}$$

A century earlier, John Wallis had introduced the numbers he called "hypergeometric numbers" and we call "factorial numbers." Euler's series is the alternating sum of those numbers.

If Euler is going to sum such a series, he first has to convince his reader, who may not be as optimistic as he is, that such series can have a meaningful value. He states his case beginning with an uncontroversial example:

Nobody doubts that the geometric series $1 + \frac{1}{2} + \frac{1}{4} + \frac{1}{8} + \frac{1}{16} + \frac{1}{32} + \text{etc.}$ converges to 2. As more terms are added, the sum approaches 2, and if 100 terms are added, the difference between the sum and 2 is a fraction with 30 digits in its denominator and a 1 in its numerator.

The series $1 + 1 + 1 + 1 + 1 + \text{etc.}$ and $1 + 2 + 3 + 4 + 5 + 6 + \text{etc.}$ whose terms do not tend toward zero, will grow to infinity and are divergent.

This is based on an idea of convergence, but since it lacks logical quantifiers, it is doomed to fall far short of modern standards of rigor.

On the other hand, Euler has interesting things to say about the alternating series of 1's. He tells us that in 1713 Leibniz said in a letter to Christian Wolff that $1 - 1 + 1 - 1 + 1 - 1 + \text{etc.}$ should have the value $\frac{1}{2}$, "based on the expansion of the fraction $\frac{1}{1+1}$." Instead of starting with a geometric series, as we usually see the calculation done today, Leibniz started with the series expansion of $\frac{1}{1-x}$, that is $1 + x + x^2 + x^3 + x^4 + \text{etc.}$ and evaluated it at $x = -1$. Likewise, Leibniz took $1 - 2 + 3 - 4 + 5 - 6 + \text{etc.}$ to be $\frac{1}{4}$ by expanding $1/(1+1)^2$. Leibniz thought that all divergent series could be evaluated.

Euler gives us four pairs of examples of divergent series:

I. $1 + 1 + 1 + 1 + 1 + 1 + \text{etc.}$

$\frac{1}{2} + \frac{2}{3} + \frac{3}{4} + \frac{4}{5} + \frac{5}{6} + \frac{6}{7} + \text{etc.}$

II. $1 - 1 + 1 - 1 + 1 - 1 + \text{etc.}$

$\frac{1}{2} - \frac{2}{3} + \frac{3}{4} - \frac{4}{5} + \frac{5}{6} - \frac{6}{7} + \text{etc.}$

III. $1 + 2 + 3 + 4 + 5 + 6 + \text{etc.}$

$1 + 2 + 4 + 8 + 16 + 32 + \text{etc.}$

IV. $1 - 2 + 3 - 4 + 5 - 6 + \text{etc.}$

$1 - 2 + 4 - 8 + 16 - 32 + \text{etc.}$

He explains that such series have been "the cause of great dissent among mathematicians of whom some deny and others affirm that such a sum can be found." He says that the first series on the list ought to be infinite, both from the nature of his understanding of infinite numbers and because, as a geometric series it has value $\frac{1}{1-1} = \frac{1}{0}$, "which is infinite."

Euler is trying to be fair in his account of this divisive controversy, and he presents the other side of the argument, writing

One could object to this argument by saying that $\frac{1}{1+a}$ is not equal to the infinite series $1 - a + a^2 - a^3 + a^4 - a^5 + a^6 -$ etc. unless a is a fraction smaller than 1, because, if we work out the division, we get

$$\frac{1}{1+a} = 1 - a + a^3 - a^3 + \cdots \pm a^n \mp \frac{a^{n+1}}{1+a},$$

and if n stands for an infinite number, then the fraction $\mp \frac{a^{n+1}}{1+a}$ cannot be omitted because it doesn't vanish unless $a < 1$, in which case the series converges."

Euler's sympathies, though, are with the side of the argument that sums even divergent geometric series, and he states what he hopes will be the final word on the issue of the existence of divergent series:

Defenders of the idea of summing divergent series resolve this paradox by devising a rather subtle means of discriminating among quantities that become negative, some that stay less than zero and others that become more than infinity. Of the first sort is -1, which by adding it to the number $a + 1$ leaves the smaller number a. Of the second sort is the -1 that arises as $1 + 2 + 4 + 8 + 16 +$ etc., which is equal to the number one gets by dividing $+1$ by -1. In the first case, the number is less than zero, and in the second case it is greater than infinity.

This can be confirmed by the following example of a sequence of fractions:

$$\frac{1}{4}, \ \frac{1}{3}, \ \frac{1}{2}, \ \frac{1}{1}, \ \frac{1}{0}, \ \frac{1}{-1}, \ \frac{1}{-2}, \ \frac{1}{-3}, \ \text{etc.}$$

where the first four terms are seen to grow, then grow to infinity, and beyond infinity they become negative. Thus the apparent absurdity is resolved in a most ingenious way.

Euler is claiming that numbers greater than infinity are the same as numbers less than zero.

Having resolved (he hopes) the question of existence, Euler turns to summing some series. He needs some ground-work. Euler asks us to consider an arbitrary (alternating) series, s:

$$s = a - b + c - d + e - f + g - h + \text{etc.}$$

Neglecting signs, the first differences are

$$b - a, \ c - b, \ d - c, \ e - d, \ \text{etc.}$$

The second differences are

$$c - 2b + a, \ d - 2c + b, \ e - 2d + c, \ \text{etc.}$$

and so forth for fourth, fifth, etc. Euler denotes the first value in each sequence of differences by a Greek letter. He takes $\alpha = b - a$, $\beta = c - 2b + a$, $\gamma = d - 3c + 3b - a$, etc., and tells us that

$$s = \frac{a}{2} - \frac{\alpha}{4} + \frac{\beta}{8} - \frac{\gamma}{16} + \frac{\delta}{32} - \text{etc.} \tag{1}$$

This is Euler's key tool for much of the rest of this paper. Euler doesn't prove his formula (1), but most readers should be able to justify his claim. My own "proof" depends on an only slightly obscure identity about binomial coefficients:

$$\sum_{k=n}^{\infty} \frac{1}{2^{k+1}} \binom{k}{n} = 1.$$

Of course, the proof is valid by modern standards of rigor only if the series s satisfies certain conditions of convergence. Still, it is a remarkable formula. If the series that defines s converges rather slowly, then this new series is likely to accelerate the convergence considerably, because the differences α, β, γ, etc. are likely not to be very large, and the formula introduces a geometric series into the denominators.

To show us how this works, Euler looks at his alternating series of 1s. For that series, $a = 1$, and all the differences, α, β, γ etc. are zero. Formula (1) gives us

$$s = \frac{1}{2} - \frac{0}{4} + \frac{0}{8} - \text{etc.}$$

$$= \frac{1}{2},$$

as promised.

If we take

$$s = 1 - 2 + 3 - 4 + 5 - 6 + \text{etc.},$$

then all the first differences are 1, and all subsequent differences are 0, so that

$$s = \frac{1}{2} - \frac{1}{4} = \frac{1}{4}.$$

If we take

$$s = 1 - 4 + 9 - 16 + 25 - 36 + \text{etc.}$$

the alternating sum of perfect squares, then our first differences are

$$3, \ 5, \ 7, \ 9, \ 11$$

and the second differences are all 2, so that

$$s = \frac{1}{2} - \frac{3}{4} + \frac{2}{8} = 0.$$

The alternating geometric sum of powers of 3 is a little bit trickier.

$$s = 1 - 3 + 9 - 27 + 81 - 243 + \text{etc.}$$

First differences are	2,	6,	18,	54,	162
Second differences are		4,	12,	36,	108
Third differences are			8,	24,	72
Fourth				16,	48
					etc.

So,

$$s = \frac{1}{2} - \frac{2}{4} + \frac{4}{8} - \frac{8}{16} + \text{etc.}$$

$$= \frac{1}{2} - \frac{1}{2} + \frac{1}{2} - \frac{1}{2} + \text{etc.}$$

$$= \frac{1}{4}.$$

In the last step, he uses his result about $1 - 1 + 1 - 1 + \text{etc.}$

Now we turn to the main problem, Wallis's hypergeometric series

$$A = 1 - 1 + 2 - 6 + 24 - 120 + 720 - 5040 + 40320 - \text{etc.}$$

Taking $1 - 1 = 0$, and dividing by 2, we get the series

$$\frac{A}{2} = 1 - 3 + 12 - 60 + 360 - 2520 + 20160 - 181440 + \text{etc.}$$

and differences

2	9	48	300	2160	17640	161280
	7	39	252	1860	15480	143640
		32	213	1608	13620	128160
			181	1395	12012	114540
				1214	10617	102528
					9403[1]	91911
						82508

This makes

$$\frac{A}{2} = \frac{1}{2} - \frac{2}{4} + \frac{7}{8} - \frac{32}{16} + \frac{181}{32} - \frac{1214}{64} + \frac{9403}{128} - \frac{82508}{256} + \text{etc.}$$

This may not look like progress, but note that the numerators are smaller than they were in the original series for A, and that there are rapidly growing denominators. There is a sense in which this series isn't "as divergent" as the original one was.

Moreover, it is still alternating, so the same trick will work again (and again, and again). Take the differences again (switching back to the series for A rather than for $A/2$ that he used in the first step) , as before, and find that they are

$$\alpha = 18/8, \quad \beta = 81/16, \quad \gamma = 456/32, \quad \delta = 3123/64, \text{ etc.}$$

Using these, we can get yet another series for A:

$$A = \frac{7}{8} - \frac{18}{32} + \frac{81}{128} - \frac{456}{512} + \frac{3123}{2048} - \frac{24894}{8192} + \text{etc.}$$

---

[1]A footnote in the *Opera Omnia* edition says that the original has a typographical error here, 9407, and that all subsequent numbers that depended on this are also in error, but are corrected in the *Opera Omnia*.

The next iteration of this process gives

$$A - \frac{5}{16} = \frac{81}{256} - \frac{132}{2048} + \frac{771}{16384} - \frac{4122}{131072} + \text{etc.}$$

so, telescoping a bit, and then neglecting the terms represented by the "etc.", he claims

$$A = \frac{5}{16} + \frac{512}{2048} + \frac{2046}{131072} + \text{etc.}$$

$$= \frac{38015}{65536}$$

$$= 0.580.$$

This is an example of what we now call an "asymptotically convergent" series. For a while, the partial sums seem to be converging, but then they swerve away from what seemed to be the limit and diverge. Modern readers may have seen them in other contexts like Euler-Maclaurin series, and they were of great interest to important 20th century mathematicians like G. H. Hardy. [H]

With some more work, not shown in the article, Euler tells us that the series can be shown to to have the value 0.59634739, but the editors of the *Opera Omnia* tell us that this last 9 should be a 6.

This is the key result of this paper, but Euler understands that some readers might not be convinced that he hasn't made any mistakes. So, he solves the same problem several other ways. For example, he finds diverging series for $1/A$ and $\log A$, and finds that similar methods also lead to a value of A near 0.59. He finds ways to write A and $1/A$ as continued fractions and evaluates those continued fractions to get still more estimates consistent with the ones before.

One of his more interesting solutions involves differential equations and infinite series. He writes

$$s = x - 1x^2 + 2x^3 - 6x^4 + 24x^5 - 120x^6 + \text{etc.}$$

and plans to evaluate it in the case $x = 1$. Differentiating gives

$$\frac{ds}{dx} = 1 - 2x + 6xx - 24x^3 + 120x^4 - \text{etc.}$$

$$= \frac{x - s}{xx}.$$

Euler loves differential equations. He rewrites this as $ds + \frac{sdx}{xx} = \frac{dx}{x}$, solves for s, and (reminding us that e is the number whose hyperbolic logarithm equals 1, that is, e is what we expect it to be) gets

$$s = e^{1/x} \int \frac{e^{-1/x}dx}{x}.$$

Substituting $x = 1$ into the definition of s gives us our alternating hyperbolic series on the left. On the right-hand side, we see that the variable x is being over-used, typical in the 18th century, and the substitution $x = 1$ should only be made outside the integral, not inside. He gets

$$1 - 1 + 2 - 6 + 12 - 120 + \text{etc.} = e \int \frac{e^{-1/x}dx}{x}$$

where the integral is taken to be between 0 and 1. Euler applies elementary numerical methods, evaluating the integrand at ten values between 0 and 1 and adding them up, and estimates that $A = 0.59637255$, consistent with his other estimates of A, but containing some small calculation errors explained in the *Opera Omnia*.

By the end of the article, Euler has estimated A at least six very different ways, and every time he gets the same estimate. When such different analyses all lead to the same conclusion, it is easy to understand why mathematicians of Euler's time believed in the utility of interesting numbers, and could believe that numbers "beyond infinity" might be negative.

References

[H]   Hardy, G. H., *Divergent Series*, Oxford University Press, New York, 1949.

[E247] Euler, Leonhard, De seriebus divergentibus, *Novi Commentarii academiae scientiarum Petropolitanae* 5, (1754/55) 1760, pp. 205–237, reprinted in *Opera Omnia* Series I vol. 14 pp. 585–617. Available online at `EulerArchive.org`.

32

Who Proved e is Irrational?

(February 2006)

Most readers will know that the constant e is, indeed irrational, even transcendental. I remember being asked to prove e was irrational on my written exams for my master's degree. It is natural, then, to ask who was the first to prove it, and to expect an easy and unambiguous answer. The answer, though, isn't as easy as we might expect, nor is it entirely unambiguous.

Here is some of what MacTutor [McT] has to say about it:

Most people accept Euler as the first to prove that e is irrational. Certainly it was Hermite who proved that e is not an algebraic number in 1873.

Note that MacTutor hedges their attribution a bit. They write "*Most* people accept Euler as the first...," (my italics) and do not commit themselves to the more definite "Euler was the first..." In this case, Euler's rival is not some earlier mathematician who might have a claim to the result, but Euler's younger protégé Johann Heinrich Lambert (1728–1777), pictured at the right. Of Lambert, MacTutor writes:

Lambert is best known, however, for his work on π. Euler had already established in 1737 that e and e^2 are both irrational. However Lambert was the first to give a rigorous proof that π is irrational. In a paper presented to the Berlin Academy in 1768 Lambert showed that, if x is a nonzero rational number, then neither e^x nor $\tan x$ can be rational.

Johann Heinrich Lambert

Note that MacTutor chooses words carefully, Euler "established," not "proved." On the other hand, Lambert's proof satisfies most standards of rigor. The hedging must be because people doubt the rigor of Euler's proof.

Our purpose in this month's column is to look at what Euler did, and to see just how rigorous Euler's results were.

Euler and Lambert both used the tools of continued fractions to produce their results. Euler's 1737 article that MacTutor mentions is "De fractionibus continuis dissertatio" [E71]. Though Euler was not the first one to study continued fractions, this article is the first comprehensive account of their properties. Euler repeats most of the elementary properties of continued fractions in the last chapter of volume 1 of his 1748 masterpiece *Introductio in analysin infinitorum* [E101]. Both of these are available in excellent English translations.

The most general form of a continued fraction is

$$a + \cfrac{\alpha}{b + \cfrac{\beta}{c + \cfrac{\gamma}{d + \cfrac{\delta}{e + \cfrac{\varepsilon}{f + \text{etc.}}}}}}$$

All the symbols, both Latin and Greek, are taken to be positive whole numbers. The Greek letters Euler calls *numerators* and the Latin are the *denominators*. In practice, the most interesting continued fractions are those for which all the numerators are 1. Continued fractions in this form are sometimes called *regular*.

It is not difficult to show that the regular continued fraction expansion of any rational number is finite, so to prove that a given number is irrational, it suffices to show that its regular expansion is not finite. We will show how this works using one of Euler's examples from the *Introductio*. We consider the number

$$\frac{e-1}{2} \approx 0.8591409142295 = \frac{8591409142295}{10000000000000}.$$

Since this number is less than 1, the first denominator, $a = 0$. Now, Euler inverts the fractional part and gets

$$\frac{10000000000000}{8591409142295} = 1 + \frac{1408590847704}{8591409142295}.$$

The next denominator is the integer part of this, so $b = 1$. Invert the fraction part of this and get

$$\frac{8591409142295}{1408590847704} = 6 + \frac{139862996071}{1408590847704}.$$

The next denominator is the integer part of this, so $c = 6$. Continue to take integer parts, and invert fractional parts, and we get

$$\frac{1408590847704}{139862996071} = 10 + \frac{9950896994}{139862996071},$$

so $d = 10$.

$$\frac{139862996071}{9950896994} = 14 + \frac{551438155}{9950896994},$$

so *e* = 14.

$$\frac{9950896994}{551438155} = 18 + \frac{25010204}{551438155},$$

so *f* = 18.

$$\frac{551438155}{25010204} = 22 + \frac{1213667}{25010204},$$

so *g* = 22.

Euler stops here, saying "If the value for *e* at the beginning had been more exact, then the sequence of quotients would have been 1, 6, 10, 14, 18, 22, 26, 34, ..., which form the terms of an arithmetic progression. It follows that

$$\frac{e-1}{2} = 0 + \cfrac{1}{1 + \cfrac{1}{6 + \cfrac{1}{10 + \cfrac{1}{14 + \cfrac{1}{18 + \cfrac{1}{22 + \text{etc.}}}}}}}\text{"}$$

Note that Euler's "arithmetic progression" doesn't start with the first denominator, but starts with the 6, after which the denominators increase by 4.

Euler adds, somewhat disingenuously, "This result can be confirmed by infinitesimal calculus."

Since the sequence of denominators clearly increases, and never terminates, this is not a finite continued fraction. Thus, by the work Euler did earlier, its value cannot be rational. Since $\frac{e-1}{2}$ is not rational, *e* cannot be rational, either.

By similar means, Euler shows that

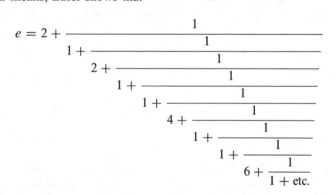

Unless Euler skipped something, the proof is done.

Alas, Euler did skip something, and he hid it in that comment, "This result can be confirmed by infinitesimal calculus." He has only *observed* that finite calculations lead to a pattern for the first few denominators, and that the pattern seems to extend indefinitely. He has not *proved* it, and he knows he has not proved it.

What could he have been thinking?

Earlier in the chapter of the *Introductio*, Euler showed how to convert a continued fraction, whether regular or not, into an alternating series. He showed that if

$$x = a + \cfrac{\alpha}{b + \cfrac{\beta}{c + \cfrac{\gamma}{d + \cfrac{\delta}{e + \cfrac{\varepsilon}{f + \text{etc.}}}}}}$$

then

$$x = a + \frac{\alpha}{b} - \frac{\alpha\beta}{b(bc + \beta)} + \frac{\alpha\beta\gamma}{(bc + \beta)(bcd + \beta d + \gamma b)} - \cdots,$$

Perhaps we can apply this alternating series formula to the coefficients we got for $\frac{e-1}{2}$ and get something related to one of the well-known series for e? But, if we do that, we get

$$x = 1 - \frac{1}{7} + \frac{1}{7 \cdot 71} - \text{etc.}$$

which does not seem related to any other well-known series for e.

So, that wasn't how he did it. If we go back to E71, we get more clues. In fact, he writes:

In the preceding sections, where I have converted the number e (whose logarithm is 1) together with its powers into continued fractions, I have only observed the arithmetic progression of the denominators and I have not been able to affirm anything except the probability of this progression continuing to infinity. Therefore, I have exerted myself in this above all: that I might inquire into the necessity of this progression and prove it rigorously. Even this goal I have pursued in a peculiar way.

Indeed his solution comes from a most surprising direction, differential equations. It lies in a form of an important differential equation called the *Ricatti equation*:

$$a \, dy + y^2 \, dx = x^{\frac{-4n}{2n+1}} \, dx.$$

Euler claims that if we substitute $p = (2n + 1)x^{\frac{1}{2n+1}}$, then the equation transforms into

$$a \, dq + q^2 \, dp = dp$$

"I have found that

$$q = \frac{a}{p} + \cfrac{1}{\cfrac{3a}{p} + \cfrac{1}{\cfrac{5a}{p} + \cfrac{1}{\cfrac{7a}{p} + \cdots + \cfrac{1}{\cfrac{(2n-1)}{p} + \cfrac{1}{x^{\frac{2n}{2n+1}} y}}}}}$$ "

This continued fraction terminates after n ratios, but if n is taken to be one of Euler's "infinite numbers," then the continued fraction goes on forever.

This is a great leap. If Euler did show this previously, then I couldn't find where he did it. The most likely places to look would be in E28 and E31, where Euler does other series analyses of the Ricatti equation, but I can't see it there, or in any other papers on differential equations that Euler wrote before he wrote E71. It was a great mystery to me.

On the other hand, the variables in the equation $a\, dq + q^2\, dp = dp$ separate to give

$$\frac{a\, dq}{1 - q^2} = dp$$

which, in turn, integrates to give

$$\frac{a}{2} \log \frac{1+q}{1-q} = p + C.$$

The constant C can be taken to be zero with the initial conditions $q = \infty$ and $p = 0$. A bit of algebra gives that

$$e^{\frac{2p}{a}} = \frac{q+1}{q-1} = 1 + \frac{1}{q-1}.$$

Just a little while ago, though, we got a continued fraction expansion for q, which we can substitute into this last expression to get

$$e^{\frac{2p}{a}} = 1 + \cfrac{1}{\cfrac{a-p}{p} + \cfrac{1}{\cfrac{3a}{p} + \cfrac{1}{\cfrac{5a}{p} + \cfrac{1}{\cfrac{7a}{p} + \text{etc.}}}}}.$$

Various values of p and a give continued fractions for various expressions involving e. For example, $p = 1$ and $a = 2$ gives

$$e = e^1 = e^{\frac{2\cdot 1}{2}}$$

$$= 1 + \cfrac{2}{1 + \cfrac{1}{6 + \cfrac{1}{10 + \cfrac{1}{14 + \text{etc.}}}}}$$

which is equivalent to the expansion for $\frac{e-1}{2}$ that Euler had observed earlier.

Euler works through a few other substitutions to derive his other observations, then writes,

> Truly everything found above follows from these formulas, by which we have expressed e and its powers as continued fractions. That is, the necessity of the progressions only observed earlier is now proved.

So, we complete the path from Euler's continued fraction solution to the Ricatti equation to the irrationality of e, but we can't be very satisfied with that solution of the differential equation. I looked at a good sample of Euler's earlier work, and can't find where Euler might have discovered this solution.

So, I was about to throw in the towel and say Euler's claim to proving the irrationality of e was kind of weak. I wasn't quite ready to let go of it, when I had one of those "right under my nose" experiences. There it was in the last five paragraphs of E71. Euler gives, in considerable detail, his proof that the continued fraction solves the Ricatti equation. We won't go into much detail; the interested reader can find the details in the Wyman and Wyman translation [E71], starting in paragraph 31. Briefly, he starts with a regular continued fraction in which the denominators form an arithmetic series. It looks like

$$s = a + \cfrac{1}{(1+n)a + \cfrac{1}{(1+2n)a + \cfrac{1}{(1+3n)a + \cfrac{1}{(1+4n)a + \text{etc.}}}}}.$$

He uses his identities from early in the paper to rewrite this as a ratio of power series involving n and a, then shows that the power series in the numerator of the ratio is related to the derivative of the power series in the denominator. This gives him a differential equation, which, three pages later, he transforms into the Ricatti equation he wanted.

It's right. It's complete, and it works. I'd been fooled when Euler suggested that he had already shown the relation between the continued fraction and the differential equation. Euler really did prove that e is irrational, and he probably regarded it as the main point of this paper.

We're ready to close this month's column. There was probably a shorter path from the question "Who proved e is irrational?" to the conclusion "Euler," but this path shows some of the details of how we learned the story. We hope you've enjoyed the adventure.

References

[C]   Chabert, Jean-Luc, et al., *A History of Algorithms: From the Pebble to the Microchip*, Translated from the French edition *Histoire d'algorithmes. Du caillou à la puce*. Editions Belin, Paris, 1994, by Chris Weeks, Springer, Berlin, 1998.

[E71]   Euler, Leonhard, "An Essay on Continued Fractions" translated by Myra F. Wyman and Bostwick F. Wyman, *Math. Systems Theory* 18 (1985) 295–328. Translation of the original De fractionibus continuis dissertatio, *Commentarii academiae scientiarum Petropolitanae* 9 (1737) 1744, pp. 98–173. Reprinted in *Opera Omnia* Series I vol. 14, pp. 187–215. Available online at EulerArchive.org.

[E101]   ——, *Introductio in analysin infinitorum*, 2 vols., Bousquet, Lausanne, 1748, reprinted in the *Opera Omnia*, Series I volumes 8 and 9. English translation by John Blanton, Springer-Verlag, 1988 and 1990. Facsimile edition by Anastaltique, Brussels, 1967.

[S]   Sandifer, Edward, "Euler and Pell," *How Euler Did It*, MAA OnLine, www.maa.org/news/howeulerdidit.html, April 2005. Reprinted in this volume on pp. 63–66.

[StA]   "The number *e*," MacTutor History of Mathematics Archive, University of St. Andrews, Scotland at www-groups.dcs.st-and.ac.uk/ history/HistTopics/e.html as revised September, 2001.

33

Infinitely Many Primes

(March 2006)

Why are there so very many prime numbers?

Euclid wondered this more than 200 years ago, and his proof that "Prime numbers are more than any assigned multitude of prime numbers" (Elements IX.20) is often considered one of the most beautiful proofs in all of mathematics.[1] Prime numbers, their detection, their frequency and their special properties remain at the heart of many of today's most exciting open questions in number theory.

This month, we return to one of Euler's early papers, *Variae observationes circa series infinitas*, to see what Euler has to say there about prime numbers. This paper has Eneström number 72, and we have already seen this paper in our column from February 2005, called "Goldbach's series." Euler wrote the paper in 1737, and it was published in 1744. Here we will find that Euler gives three answers to the question "How many prime numbers are there?" and, in a way, helps us understand why there are so many of them.

Since we discussed the first part of this paper last February, we will start in the middle of this paper, with Euler's Theorem 7:

The product continued to infinity of this fraction

$$\frac{2 \cdot 3 \cdot 5 \cdot 7 \cdot 11 \cdot 13 \cdot 17 \cdot 19 \cdot etc.}{1 \cdot 2 \cdot 4 \cdot 6 \cdot 10 \cdot 12 \cdot 16 \cdot 18 \cdot etc.}$$

in which the numerators are prime numbers and the denominators are one less than the numerators, equals the sum of this infinite series

$$1 + \frac{1}{2} + \frac{1}{3} + \frac{1}{4} + \frac{1}{5} + \frac{1}{6} + etc.,$$

and they are both infinite.

The factors of the infinite product can be rewritten as

$$\frac{p}{p-1} = \frac{1}{1 - 1/p},$$

---

[1] It was voted number 3 in a 1988 survey by *Mathematical Intelligencer*. See [W, p. 126]

so, in modern notation, this theorem can be rewritten as

$$\sum_{k=1}^{\infty} \frac{1}{k} = \prod_{p} \frac{1}{1 - 1/p}.$$

Some readers will recognize this as the sum-product formula for the Riemann zeta function at the value $s = 1$. Others may recognize it as part of the cover illustration of William Dunham's wonderful book on Euler and his work. [D] Euler offers the following proof of this theorem.

Proof. Let

$$x = 1 + \frac{1}{2} + \frac{1}{3} + \frac{1}{4} + \frac{1}{5} + \frac{1}{6} + \text{etc.}$$

Then

$$\frac{1}{2}x = \frac{1}{2} + \frac{1}{4} + \frac{1}{6} + \frac{1}{8} + \text{etc.}$$

This leaves

$$\frac{1}{2}x = 1 + \frac{1}{3} + \frac{1}{5} + \frac{1}{7} + \text{etc.}$$

Note that there are no even numbers left in the denominators on the right-hand side. Now, to eliminate the denominators that are divisible by 3, we divide both sides by 3, to get

$$\frac{1}{2} \cdot \frac{1}{3}x = \frac{1}{3} + \frac{1}{9} + \frac{1}{15} + \frac{1}{21} + \text{etc.}$$

Subtracting again eliminates all remaining denominators that are multiples of 3, leaving

$$\frac{1}{2} \cdot \frac{2}{3}x = 1 + \frac{1}{5} + \frac{1}{7} + \frac{1}{11} + \frac{1}{13} + \text{etc.}$$

This process is like the ancient sieve of Eratosthenes because, at each stage, it eliminates a prime denominator and all remaining multiples of that prime denominator. Eventually, everything on the right will be eliminated except the first term, 1.

Euler carries us through one more iteration of his process. He divides this last equation by 5, and does a small rearrangement of the way he writes the product of fractions, to get

$$\frac{1 \cdot 2}{2 \cdot 3} \cdot \frac{1}{5}x = \frac{1}{5} + \frac{1}{25} + \frac{1}{35} + \text{etc.}$$

Subtracting leaves

$$\frac{1 \cdot 2 \cdot 4}{2 \cdot 3 \cdot 5}x = 1 + \frac{1}{7} + \frac{1}{11} + \frac{1}{13} + \text{etc.}$$

In the same way, terms with denominators that are multiples of 7, 11, and so forth for all prime numbers, will be eliminated, leaving

$$\frac{1 \cdot 2 \cdot 4 \cdot 6 \cdot 10 \cdot 12 \cdot 16 \cdot 18 \cdot 22 \cdot \text{etc.}}{2 \cdot 3 \cdot 5 \cdot 7 \cdot 11 \cdot 13 \cdot 17 \cdot 19 \cdot 21 \cdot \text{etc.}}x = 1.$$

But, since x is already known to be the sum of the harmonic series, the desired result is immediate.   Q.E.D.

From its very first step, this proof does not satisfy modern standards of rigor, but it isn't too much work for a modern mathematician to recast the theorem to say that the harmonic series has a finite limit if and only if the infinite product does, and then to begin the proof "suppose that the harmonic series converges to a value x."

Nonetheless, if we accept the result, then we have a short proof that there are infinitely many primes. For the product

$$\frac{2 \cdot 3 \cdot 5 \cdot 7 \cdot 11 \cdot 13 \cdot 17 \cdot 19 \cdot \text{etc.}}{1 \cdot 2 \cdot 4 \cdot 6 \cdot 10 \cdot 12 \cdot 16 \cdot 18 \cdot \text{etc.}}$$

to diverge it must be an *infinite* product, hence there must be infinitely many prime numbers.

Though it isn't exactly relevant to our topic, the next theorem, Theorem 8 is extremely important:

Theorem 8. *If we use the series of prime numbers to form the expression*

$$\frac{2^n}{(2^n - 1)} \cdot \frac{3^n}{(3^n - 1)} \cdot \frac{5^n}{(5^n - 1)} \cdot \frac{7^n}{(7^n - 1)} \cdot \frac{11^n}{(11^n - 1)} \cdot etc.$$

then its value is equal to the sum of this series

$$1 + \frac{1}{2^n} + \frac{1}{3^n} + \frac{1}{4^n} + \frac{1}{5^n} + \frac{1}{6^n} + \frac{1}{7^n} + etc.$$

In modern notation, this is the familiar sum-product formula for the Riemann zeta function:

$$\zeta(s) = \sum_{k=1}^{\infty} \frac{1}{k^s} = \prod_{p \text{ prime}} \frac{1}{1 - 1/p^s},$$

and Euler's proof of Theorem 8 is almost exactly like the proof of Theorem 7, with the exponent n included. Also, the proof of Theorem 8 is correct by modern standards because all the series involved are absolutely convergent.

Occasionally, someone will insist that, because Euler proved this Theorem 8, the function we call the Riemann zeta function ought to be called the *Euler* zeta function. We disagree. Euler's zeta function is a function of a real variable, and Euler never treats its value except for positive integers. Riemann extended the function to *complex* values, where its most interesting and important properties are found. If you want to, you can try to say the Euler zeta function is a function of a real variable, but the Riemann zeta function is complex. The forces of history probably won't let you though, and that is fair.

We jump forward to Theorem 19 for Euler's second result about how many prime numbers there are. His theorem is:

Theorem 19. *The sum of the series of reciprocals of prime numbers*

$$\frac{1}{2} + \frac{1}{3} + \frac{1}{5} + \frac{1}{7} + \frac{1}{11} + \frac{1}{13} + etc.$$

is infinitely large, and is infinitely less than the harmonic series,

$$1 + \frac{1}{2} + \frac{1}{3} + \frac{1}{4} + \frac{1}{5} + etc.$$

Moreover, the first sum is almost the logarithm of the second sum.[2]

Euler's proof of this again relies on adding and subtracting series that do not converge, and so does not meet modern standards of rigor. Unfortunately, Euler's proof can't really be "repaired." This time, we omit the proof.

Note that the first statement in the proof, that the series of reciprocals of primes diverges, not only tells us that there are infinitely many primes (since a finite series cannot diverge), but tells us that the primes are "dense" enough that the sum of their reciprocals diverges. The sum of the reciprocals of the square numbers, on the other hand, converges to $\pi^2/6$, as Euler showed a few years earlier when he solved the Basel problem. In this sense, there are "more" prime numbers than there are square numbers.

Finally, we come to Euler's third measure of how many primes there are. This comes from the second part of Theorem 19. This will require some interpretation and some analysis, but we hope to show that this remark is closely related to the so-called Prime Number Theorem.

Earlier in this paper, Euler made a cryptic remark about the "value" of the harmonic series, "if the absolute infinity is taken to be $= \infty$ then this expression will have the a value $= \ln \infty$, which is the smallest of all the infinite powers."

The part about "$= \ln \infty$" is fairly easy to interpret. In modern terms, it probably means that, for large values of n,

$$\sum_{k=1}^{n} \frac{1}{k} \approx \ln n.$$

The part about "smallest of all the infinite powers" isn't as clear, but I think it is about what we now call p-series, that the smallest value of p for which the p-series

$$\sum_{k=1}^{\infty} \frac{1}{k^p}$$

diverges is $p = -1$, the case of the harmonic series. Euler describes this, rather cryptically, saying that the harmonic series is the smallest series that sums to infinity, and that, therefore, the infinity to which it sums is the smallest of all infinities. Few people today would accept this contention.

Interpreting the rest of Euler's claim in the same way, it would seem to translate into the approximation

$$\sum_{\substack{p \text{ prime} \\ p < n}} \frac{1}{p} \approx \ln(\ln n).$$

We can find this as a theorem in modern number theory books, for example theorem 427 in [HW].

We can differentiate this to find out how much the sum is expected to increase if we proceed from n to $n + 1$. We find that the derivative is $\frac{1}{n \ln(n)}$. We can apply some ideas from probability. If n is prime, then the sum will increase by $\frac{1}{n}$, and if n is not prime, then the sum will not increase. Hence, the "probability" that n is prime is about $\frac{1}{\ln n}$.

---

[2]In Euler's Latin: *Atque illius summa est huius summae quasi logarithmus.*

According to MacTutor [McT] "The statement that the density of primes is $1/\log n$ is known as the *Prime Number Theorem*." Moreover, Legendre observed this fact about the density of primes in 1798, and Gauss claimed to have observed it in 1793, but it was not proved until 1896 when Hadamard and de la Vallée Poussin independently discovered proofs. Precedence is usually given to Gauss's observation.

However, as we have just seen, the Prime Number Theorem is an easy consequence of Euler's Theorem 19.

Euler scooped Gauss by more than fifty years.

References

[E]   Euclid, *Elements*, Sir Thomas Heath, tr., 3 vol., Dover, 1956.

[E72]   Euler, Leonhard, Variae observationes circa series infinitas, *Commentarii academiae scientiarum Petropolitanae* 9 (1737), 1744, pp. 160–188. Reprinted in *Opera Omnia* Series I volume 14, pp. 216–244. Also available on line at www.EulerArchive.org.

[HW]   Hardy, G. H, and E. M. Wright, *An Introduction to the Theory of Numbers*, 5th ed., Oxford Univ. Press, 1979.

[S]   Sandifer, Ed, "Goldbach's series," *How Euler Did It*, MAA OnLine, www.maa.org/news/howeulerdidit.html, February 2005. Reprinted in this volume on pp. 167–170.

[D]   Dunham, William, *Euler: The Master of Us All*, Dolciani Mathematical Expositions vol. 22, Mathematical Association of America, Washington, DC, 1999.

[McT]   "Prime Numbers," MacTutor History of Mathematics Archive, www-groups.dcs.st-and.ac.uk/~history/HistTopics/Prime_numbers.html.

[W]   Wells, David, *The Penguin Book of Curious and Interesting Mathematics*, Penguin, London, 1997.

34

Formal Sums and Products

(July 2006)

In June 2006 at our MAA Section meeting, George Andrews gave a nice talk about the delicate and beautiful relations among infinite sums, infinite products and partitions. Dr. Andrews, the Evan Pugh Professor of Mathematics at Penn State, is known even among non-mathematicians for his 1976 discovery of "Ramanujan's Lost Notebook," a collection of 138 pages of notes that had lain unnoticed in the archives at Trinity College, Cambridge. A nice account of the colorful history of the "Lost Notebook" is online at [B].

Dr. Andrews described some of the tools he uses to understand and to extend Ramanujan's work. They include functions called *q-series*, [W] defined and denoted by

$$(a;q)_n = \prod_{k=0}^{n-1} \left(1 - aq^k\right).$$

One *q*-series is particularly important, and is known among friends of Ramanujan and Andrews as Euler's function:

$$\phi(q) = (q;q)_\infty = \prod_{k=1}^{\infty} \left(1 - q^k\right).$$

As we will see (and as long-time readers of this column have already seen in [S2005],) when these products are expanded into a sum, then the coefficients of the resulting series contain information about partitions.

This month's column will discuss some of the tools Euler developed to understand the relations among infinite products, infinite sums and combinatorics that make the work of Andrews and Ramanujan so beautiful.

Euler danced among products, sums and combinations several times. In [E19] he used products to discover the gamma function. In [E41], he linked products and sums to solve the Basel problem. Then in [E158] he linked sums, products and partitions to solve Philip Naudé's problem. He touched on the connections several other times, but rather than trace how he developed the ideas, we will look at his unified presentation in Chapter 15 of the *Introductio in analysin infinitorum* [E101], the chapter titled "On series that arise from the expansion of products."

Let us recall when we first learned about quadratic equations. We learned that if a product, $(x - r)(x - s)$ expands into a sum, $x^2 + bx + c$, then the roots of the quadratic are r and s, that $c = rs$ and that $b = -(r + s)$.

Euler learned from a slightly different book. He learned this same fact using a product of the form $(1 + \alpha z)(1 + \beta z)$ and a quadratic $1 + Az + Bz^2$. Though this makes the roots a little harder to write, it makes the sums and products a little easier, so that $A = \alpha + \beta$ and $B = \alpha\beta$.

Euler opens Chapter 15 by generalizing this result to many factors. He asks us to consider the sum and product

$$1 + Az + Bz^2 + Cz^3 + \text{etc.} = (1 - \alpha z)(1 - \beta z)(1 - \gamma z) \text{ etc.} \tag{1}$$

Then he explains that

$$A = \alpha + \beta + \gamma + \delta + \varepsilon + \zeta + \text{etc.}$$
$$= \text{ the sum of the coefficients taken individually,}$$

$$B = \alpha\beta + \alpha\gamma + \beta\gamma + \alpha\delta + \beta\delta + \gamma\delta + \text{etc.}$$
$$= \text{ the sum of the products taken two at a time,}$$

$$C = \text{ the sum of the products taken three at a time,}$$

$$D = \text{ the sum of the products taken four at a time,}$$
$$\text{etc.}$$

Euler is deliberately vague about whether he means these sums and products to be finite or infinite. This is partly because he, like most others of his time, believed in something called the "Principle of Continuity" or the "Principle of Continuation." This was a philosophical principle of Leibniz, stating, roughly, that if two things are not much different, then their effects and properties will not be much different, either. [P] Leibniz captured the principle with his aphorism "Nature makes no leaps." Thus, the relations between factors and coefficients that are true for finite sums and products should also be true for infinite ones.

Euler also ignored all issues of convergence in this discussion, unlike today when we wave our hands and either mumble "formal power series" or "suitable radius of convergence."

For his first example, Euler takes the Greek letters α, β, γ, δ, etc., to be the sequence of prime numbers, 2, 3, 5, 7, 11, 13, etc. Now we'll do in three or four steps what Euler does in one step.

Consider the product

$$(1 + 2z)(1 + 3z)(1 + 5z)(1 + 7z)(1 + 11z) \text{ etc.}$$

Equation (1) and the properties of A, B, C, etc. tell us that this product expands to give

$$1 + (2 + 3 + 5 + 7 + 11 + 13 + \text{etc.})z$$
$$+ (2 \cdot 3 + 2 \cdot 5 + 3 \cdot 5 + 2 \cdot 7 + 3 \cdot 7 + 5 \cdot 7 + \text{etc.})z^2$$
$$+ (2 \cdot 3 \cdot 5 + 2 \cdot 3 \cdot 7 + 3 \cdot 5 \cdot 7 + 2 \cdot 3 \cdot 11 + \text{etc.})z^3$$
$$+ (\text{etc.})z^4 + \text{etc.}$$

Euler takes $z = 1$, performs the multiplications, rearranges the terms into increasing order, and names the whole sum P:

$$P = 1 + 2 + 3 + 5 + 6 + 7 + 10 + 11 + 13 + 14 + 15 + 17 + \text{etc.}$$

"in which series," he tells us, "all natural numbers occur except those that are powers, and those which are divisible by some power." We would call this the series of "square-free" numbers, those that are not divisible by any square number except 1.

Note that Euler isn't really interested in the *value* of this series; rather he wants us to see the numbers that occur in the series itself.

For his next example, he takes his coefficients α, β, γ, δ, etc. to be "any power of the prime numbers." His formula makes it clear that he means the power to be negative, as he writes

$$P = \left(1 + \frac{1}{2^n}\right)\left(1 + \frac{1}{3^n}\right)\left(1 + \frac{1}{5^n}\right)\left(1 + \frac{1}{7^n}\right)\left(1 + \frac{1}{11^n}\right) \text{ etc.}$$

In many ways, this is a lot like his previous example. This formula expands into a series to give

$$P = 1 + \frac{1}{2^n} + \frac{1}{3^n} + \frac{1}{5^n} + \frac{1}{6^n} + \frac{1}{7^n} + \frac{1}{10^n} + \frac{1}{11^n} + \text{etc.}$$

where again the denominators are based on square-free numbers. It is a more interesting example than it seems to be. First, it is true. For $n > 1$, both the series and the product converge, and they converge to the same (finite) value.

Second, it is an example of the Principle of Continuation applied in reverse. We had a fact about infinite series that do not converge. The argument that convinced the reader (maybe) of that fact also applies to the closely related infinite series that *do* converge. If the first argument convinced us, then this argument should convince us, as well.

Third, Euler has a plan. These examples really are going somewhere. He's not telling us where, though.

His next example has a new idea, but no clues where he's going. Take the negatives of the powers in the previous example, so that

$$P = \left(1 - \frac{1}{2^n}\right)\left(1 - \frac{1}{3^n}\right)\left(1 - \frac{1}{5^n}\right)\left(1 - \frac{1}{7^n}\right)\left(1 - \frac{1}{11^n}\right) \text{ etc.} \tag{2}$$

The formula tells us that this expands to give

$$P = 1 - \frac{1}{2^n} - \frac{1}{3^n} - \frac{1}{5^n} + \frac{1}{6^n} - \frac{1}{7^n} + \frac{1}{10^n} - \frac{1}{11^n} + \text{etc.} \tag{3}$$

where the denominators again include only the square-free bases, but the signs are determined by the number of prime divisors in the denominator. Those with an odd number of prime divisors, like the primes themselves, or like $30 = 2 \cdot 3 \cdot 5$, have the negative sign, but those with an even number of divisors, like $6 = 2 \cdot 3$, $10 = 2 \cdot 5$ and $15 = 3 \cdot 5$, have a positive sign.

Remember these formulas. We will see them again later.

Now we turn to quotients. Euler asks us to "now consider this expression"

$$\frac{1}{(1-\alpha z)(1-\beta z)(1-\gamma z)(1-z\delta)(1-\varepsilon z)\text{etc.}}.$$

Again he expands the quotient into a series

$$1 + Az + Bz^2 + Cz^3 + Dz^4 + Ez^5 + Fz^6 + \text{etc.}$$

and, without explanation, tells us that

> $A =$ the sum of the coefficients taken one at a time,
>
> $B =$ the sum of the coefficients multiplied together two at a time, possibly with repetition,
>
> $C =$ *the sum of the coefficients multiplied together two at a time, again with repetition allowed,*
>
> $D =$ the sum of products taken four at a time,
>
> etc.

Let's fill in a few steps that Euler omitted, and give a simple example. Suppose there are only two factors in the denominator, so our quotient is

$$\frac{1}{(1-\alpha z)(1-\beta z)} = \frac{1}{(1-\alpha z)} \cdot \frac{1}{(1-\beta z)}.$$

Each of the factors on the right expands into a geometric series;

$$\frac{1}{1-\alpha z} = 1 + \alpha z + \alpha^2 z^2 + \alpha^3 z^3 + \alpha^4 z^4 + \text{etc.},$$

and likewise for the other factor,

$$\frac{1}{1-\beta z} = 1 + \beta z + \beta^2 z^2 + \beta^3 z^3 + \beta^4 z^4 + \text{etc.}$$

We note in passing that Euler gets these geometric series identities by "actual division" of the quotient, and not from manipulations on the series. In other words, he finds the series given the quotient, instead of doing what we usually do today, start with the series and find its analytical value.

When we multiply together these two series, we get

$$\frac{1}{(1-\alpha z)(1-\beta z)} = 1 + (\alpha + \beta)z + (\alpha\alpha + \alpha\beta + \beta\beta)z^2 + (\alpha^3 + \alpha^2\beta + \alpha\beta^2 + \beta^3)z^3 + \text{etc.}$$

We see that the coefficients A, B, C, etc., are as Euler described.

Euler's next example is deceptively simple. He takes a quotient with just one factor, $\alpha = 1/2$ and sets $z = 1$, to get

$$\frac{1}{1-\frac{1}{2}} = 1 + \frac{1}{2} + \frac{1}{4} + \frac{1}{8} + \frac{1}{16} + \frac{1}{32} + \text{etc.}$$

Note that Euler is thinking of this as expanding the quotient, not as summing the series, and that he's interested in the terms of the series, not the value of the sum.

Euler proceeds to the case $\alpha = \frac{1}{2}$, $\beta = \frac{1}{3}$ and claims

$$\frac{1}{(1-\frac{1}{2})(1-\frac{1}{3})} = 1 + \frac{1}{2} + \frac{1}{3} + \frac{1}{4} + \frac{1}{6} + \frac{1}{8} + \frac{1}{9} + \frac{1}{12} + \frac{1}{16} + \frac{1}{18} + \text{etc.},$$

where the denominators involve only numbers with no prime divisors other than 2 and 3.

This begs to be extended, so Euler takes α, β, γ, δ, etc., to be the sequence of prime numbers, 2, 3, 5, 7, 11, 13, etc., and $z = 1$. Then he writes (by the Principle of Continuation) that if

$$P = \frac{1}{(1-\frac{1}{2})(1-\frac{1}{3})(1-\frac{1}{5})(1-\frac{1}{7})(1-\frac{1}{11})(1-\frac{1}{13})} \text{ etc.}$$

then

$$P = 1 + \frac{1}{2} + \frac{1}{3} + \frac{1}{4} + \frac{1}{5} + \frac{1}{6} + \frac{1}{7} + \frac{1}{8} + \frac{1}{9} + \text{etc.}$$

This last is the harmonic series, a series we know diverges. Euler would say that its value is $\ln \infty$. If we write this in modern notation, using Sigma for sums and Pi for products, we get

$$\sum_{k=1}^{\infty} \frac{1}{k} = \prod_{p \text{ prime}} \frac{1}{1 - \frac{1}{p}},$$

the famous Sum-Product formula for the Riemann zeta function. This formula has a prominent place on the cover of William Dunham's book *Euler: The Master of Us All* [D]. Still, the formula isn't really *true*. Since the harmonic series diverges, it can't be said to have a real value, so its value can't really be equal to anything else.

Again, the Principle of Continuation has something to add. Euler goes on to tell us that if

$$P = \frac{1}{\left(1-\frac{1}{2^n}\right)\left(1-\frac{1}{3^n}\right)\left(1-\frac{1}{5^n}\right)\left(1-\frac{1}{7^n}\right)\left(1-\frac{1}{11^n}\right)\left(1-\frac{1}{13^n}\right)} \text{ etc.},$$

then, by the same calculation,

$$P = 1 + \frac{1}{2^n} + \frac{1}{3^n} + \frac{1}{4^n} + \frac{1}{5^n} + \frac{1}{6^n} + \frac{1}{7^n} + \frac{1}{8^n} + \frac{1}{9^n} + \text{etc.}$$

Again, in modern notation, this is

$$\sum_{k=1}^{\infty} \frac{1}{k^n} = \prod_{p \text{ prime}} \frac{1}{1 - \frac{1}{p^n}}$$

and this time, for $n > 1$, both the sum and the product converge, and they are equal. The sum on the left is Riemann's zeta function, and this fact is one of the fundamental properties of the zeta function.

This is the second time Euler proved this formula. We saw the first proof a few months ago [S2006] when we were looking at some results from [E72]. This proof is quite different.

This chapter of the *Introductio* goes on quite a bit farther, using these techniques to calculate values of particular series, but we will wrap it up with one last result, a lesser-known property of the zeta function.

As before, take

$$P = \cfrac{1}{\left(1 - \frac{1}{2^n}\right)\left(1 - \frac{1}{3^n}\right)\left(1 - \frac{1}{5^n}\right)\left(1 - \frac{1}{7^n}\right)\left(1 - \frac{1}{11^n}\right)\left(1 - \frac{1}{13^n}\right)} \text{ etc.}$$

so that P is the zeta function

$$P = 1 + \frac{1}{2^n} + \frac{1}{3^n} + \frac{1}{4^n} + \frac{1}{5^n} + \frac{1}{6^n} + \frac{1}{7^n} + \frac{1}{8^n} + \frac{1}{9^n} + \text{ etc.}$$

Now, take Q to be the reciprocal of P, so that

$$Q = \cfrac{1}{\left(1 - \frac{1}{2^n}\right)\left(1 - \frac{1}{3^n}\right)\left(1 - \frac{1}{5^n}\right)\left(1 - \frac{1}{7^n}\right)\left(1 - \frac{1}{11^n}\right)\left(1 - \frac{1}{13^n}\right)} \text{ etc.}$$

This is the same product we saw in formula (2), (did you remember it as we told you to?) except it's named Q now instead of P. So, we know from formulas (2) and (3) that this product expands as a series to give

$$Q = 1 - \frac{1}{2^n} - \frac{1}{3^n} - \frac{1}{5^n} + \frac{1}{6^n} - \frac{1}{7^n} + \frac{1}{10^n} - \frac{1}{11^n} + \text{ etc.}$$

From the product forms, it is completely obvious that $PQ = 1$, but as a series, the fact is quite remarkable.

There is always something interesting to learn from the *Introductio*. Ramanujan appreciated this kind of analysis, and people like George Andrews remind us that we still have much to learn by following the footsteps of Euler and Ramanujan.

References

[B]   Berndt, Bruce C., "The remaining 40% of Ramanujan's lost notebook," online at www.math.uiuc.edu/ berndt/articles/kyoto.pdf.

[D]   Dunham, William, *Euler: The Master of Us All*, MAA, Washington, DC, 1999.

[E19]   Euler, Leonhard, De progressionibus transcendentibus seu quarum termini generales algebraice dari nequeunt, *Commentarii academiae scientiarum Petropolitanae* 5 (1730/31), 1738, pp. 36–57. Reprinted in *Opera Omnia* Series I vol. 14, pp.1–24. A copy of the original as well as a translation by Stacy Langton are available online at www.EulerArchive.org.

[E41]   ——, De summis serierum reciprocarum, *Commentarii academiae scientiarum Petropolitanae* 7 (1734/35), 1740, pp. 123–134. Reprinted in *Opera Omnia* Series I volume 14, pp.73–86. A copy of the original as well as a translation by Jordan Bell are available online at www.EulerArchive.org.

[E72] ——, Variae observationes circa series infinitas, *Commentarii academiae scientiarum Petropolitanae* 9 (1737), 1744, pp. 160–188. Reprinted in *Opera Omnia* Series I volume 14, pp. 216–244. Available online at www.EulerArchive.org.

[E101] ——, *Introductio in analysin infinitorum*, 2 vols., Bousquet, Lausanne, 1748, reprinted in the *Opera Omnia*, Series I volumes 8 and 9. English translation by John Blanton, Springer-Verlag, 1988 and 1990. Facsimile edition by Anastaltique, Brussels, 1967.

[E158] ——, Observationes analyticae variae de combinationibus, *Commentarii academiae scientiarum Petropolitanae* 13 (1741/43) 1751, pp. 64–93, reprinted in *Opera Omnia* Series I vol. 2 pp. 163–193. Available online at www.EulerArchive.org.

[P] Philosophy Professor, "Law or principle of continuity," www.philosophyprofessor. com / philosophies/continuity-law-or-principle.php, June 12, 2006.

[S2005] Sandifer, Ed, "Philip Naudé's problem," *How Euler Did It*, MAA Online, October 2005. Reprinted in this volume on pages 85–90.

[S2006] ——, "Infinitely many primes," *How Euler Did It*, MAA Online, March 2006. Reprinted in this volume on pages 191–195.

[W] "Q-series," Wikipedia, online at en.wikipedia.org/wiki/Q-series. June 12, 2006.

35

Estimating the Basel Problem

(December, 2003)

In the lives of famous people, we can often identify the first thing they did that made them famous. For Thomas Edison, it was probably his invention of the phonograph in 1877. Abraham Lincoln first made his name in the Lincoln-Douglas Debates of 1858, though Steven A. Douglas won that election for the U. S. Senate, not Abraham Lincoln.

Leonhard Euler's first celebrated achievement was his solution in 1735 of the "Basel Problem", finding an exact value of the sum of the squares of the reciprocals of the integers, that is

$$1 + \frac{1}{4} + \frac{1}{9} + \frac{1}{16} + \frac{1}{25} + \frac{1}{36} + \frac{1}{49} + \cdots .$$

Bill Dunham [D] gives a wonderful account of Euler's solution in his book *Euler The Master of Us All*, published by the MAA in 1999. However, just as Edison's invention of the phonograph depended critically on his invention of waxed paper a few years earlier, so also Euler's solution to the Basel Problem had its roots in a result from 1730 on estimating integrals.

Pietro Mengoli (1625–1686) posed the Basel problem in 1644. The problem became well known when Jakob Bernoulli wrote about it in 1689. Jakob was the brother of Euler's teacher and mentor Johann Bernoulli, who probably showed the problem to Euler. By the 1730s, the problem had thwarted many of the day's best mathematicians, and it had achieved the same kind of mystique that Fermat's Last Theorem had before 1993.

In 1730 Euler is interested in problems he calls "interpolation of series." Given a process defined for whole numbers, he seeks meaningful ways to extend the definitions of those processes to non-integer values. For example, he already extended what he called the hypergeometric series and we call the factorial function, $n! = 1 \cdot 2 \cdot 3 \cdots n$ to give a definition that worked for fractional values. The function he devised is now called the gamma function.

In the paper that bears the index number E20, Euler does the same thing for the partial sums of the harmonic series, $1 + \frac{1}{2} + \frac{1}{3} + \frac{1}{4} + \cdots$. If the first partial sum of this series is 1 and the second is $\frac{3}{2}$ and the third is $\frac{11}{6}$, and so forth, Euler asks what value might be assigned to the $\frac{1}{2}$th or the $\frac{3}{2}$th partial sums. Euler's answer lies in integration and geometric series.

First, Euler recalls the formula for the sum of a finite geometric series,

$$1 + x + x^2 + x^3 + \cdots + x^{n-1} = \frac{1 - x^n}{1 - x}.$$

Here, n is the number of terms, and the formula can be applied for any value of n, even though it may not be clear what the formula might mean if n is not a whole number.

Now, Euler integrates both sides. On the left, he gets

$$x + \frac{x^2}{2} + \frac{x^3}{3} + \frac{x^4}{4} + \cdots + \frac{x^n}{n}.$$

Taking $x = 1$ gives the desired nth partial sum of the harmonic series. On the right, he gets

$$\int \frac{1 - x^n}{1 - x} \, dx.$$

Our modern notation for definite integrals had not yet evolved, so instead of writing

$$\int_0^1 \frac{1 - x^n}{1 - x} \, dx,$$

Euler adds that we should "take it $= 0$ if $x = 0$, and then set $x = 1$." So Euler uses this integral as the value for the nth partial sum of the harmonic series, even if n is not a whole number.

For most values of n, the integral

$$\int_0^1 \frac{1 - x^n}{1 - x} \, dx$$

is rather difficult to evaluate. In the case $n = \frac{1}{2}$, though, it is not too difficult to find that

$$\int_0^1 \frac{1 - \sqrt{x}}{1 - x} \, dx = \int_0^1 \frac{1}{1 + \sqrt{x}} \, dx = 2 - 2\ln 2,$$

thus giving a value to the $\frac{1}{2}$th partial sum.

Euler can always find more to do with a new technique. This integration idea works because integration gives a relation between geometric and harmonic series. In particular, (overlooking, as Euler did, the constant of integration)

$$\int \frac{1 - x^n}{1 - x} \, dx = \int \left(1 + x + x^2 + \cdots + x^{n-1}\right) dx = x + \frac{x}{2} + \frac{x^2}{3} + \cdots + \frac{x^n}{n}.$$

If we integrate again, then on the right-hand side we get

$$\frac{x^2}{1 \cdot 2} + \frac{x^3}{2 \cdot 3} + \frac{x^4}{3 \cdot 4} + \cdots + \frac{x^{n+1}}{n \cdot (n + 1)}.$$

Here, the denominators are products of consecutive integers. These are tantalizingly like, but are not exactly, squares. The Basel Problem seems to be lurking nearby.

Euler, though, sees a way to get the perfect squares in the denominators that he wants so badly. He divides by x before integrating again. He sees that

$$\int \frac{1}{x} \left(x + \frac{x}{2} + \frac{x^2}{3} + \cdots + \frac{x^n}{n} \right) dx = x + \frac{x^2}{2 \cdot 2} + \frac{x^3}{3 \cdot 3} + \cdots + \frac{x^n}{n \cdot n},$$

again, neglecting the constant of integration. Euler doesn't have a very good notation for double integrals, either, so he writes this in a way that we would now regard as nonsense:

$$\int \frac{dx}{x} \int \frac{1 - x^n}{1 - x} dx = 1 + \frac{1}{4} + \frac{1}{9} + \cdots + \frac{1}{n^2}$$

Euler does give us instructions about how to evaluate this, and they translate into modern notation to be

$$\int_0^1 \frac{1}{x} \left(\int_0^x \frac{1 - y^n}{1 - y} dy \right) dx.$$

To solve the Basel Problem, all Euler has to do is evaluate this integral and then take its limit as n goes to infinity. This does not seem like progress, because this integral is very difficult to evaluate. However, Euler uses a very clever approximation technique to get a numerical value, 1.644924, for this integral. The series itself converges rather slowly, and you would have to sum more than 30,000 terms to get this degree of accuracy directly.

This value does not mean much to most people, but among his other skills, Euler is a fantastic calculator. He sees this value is close to $\pi^2/6$. Armed with this hint, he begins to attack the Basel Problem using series for trigonometric functions, in particular, the Taylor Series for $\sin x/x$. In fact, Euler was the first one to call them "Taylor's series." Five years later, he publishes his solution and the world of mathematics knows his name.

References

[D] Dunham, William, *Euler: The Master of Us All*, MAA, Washington, DC, 1999.

36

Basel Problem with Integrals

(March 2004)

Euler's brilliant 1734 solution to the Basel problem, to find the value of the series

$$1 + \frac{1}{4} + \frac{1}{9} + \frac{1}{16} + \frac{1}{25} + \frac{1}{36} + \text{etc.}$$

[E41] brought him great fame, but it also depended on some assumptions that are rather difficult to justify. In particular, at a key point in the solution, Euler notes that the function $\frac{\sin x}{x}$ and the infinite product

$$\left(1 + \frac{x}{\pi}\right)\left(1 - \frac{x}{\pi}\right)\left(1 + \frac{x}{2\pi}\right)\left(1 - \frac{x}{2\pi}\right)\left(1 + \frac{x}{3\pi}\right)\left(1 - \frac{x}{3\pi}\right)\cdots$$

have exactly the same roots and have the same value at $x = 0$, so Euler asserts that they describe the same function. Euler is correct that they describe the same function, but these reasons are insufficient to guarantee it. For example, the function $e^x \frac{\sin x}{x}$ also has the same roots and the same value at $x = 0$, but it is a different function. This seems to be a modern objection, not raised in Euler's time.

Nonetheless, Euler seemed to understand that there was something mysterious or incomplete in his explanation of this step. He wrote some other papers, for example E61, in which he tried to extend and justify this infinite product technique, but he never got very far with clearing the fog out of the solution.

It is generally assumed that was where Euler left the issue. However, in 1741, he wrote a seldom-read paper in French, published in a rather obscure literary journal, in which he gives a completely different solution to the Basel problem, one that does not depend on the mysteries of infinite products.

Euler begins by asking us to consider a circle of radius 1. He takes s to be arc length, and takes $x = \sin s$, or, equivalently, $s = \arcsin x$. Then, working with differentials as he always does,

$$ds = \frac{dx}{\sqrt{1 - xx}} \quad \text{and} \quad s = \int \frac{dx}{\sqrt{1 - xx}}.$$

Now, he multiplies these together to get

$$s \, ds = \frac{dx}{\sqrt{1-xx}} \int \frac{dx}{\sqrt{1-xx}}.$$

He integrates both sides from $x = 0$ to $x = 1$. On the left, the antiderivative is $\frac{ss}{2}$, and, as x goes from 0 to 1, s goes from 0 to $\frac{\pi}{2}$, so he gets on the left $\frac{\pi\pi}{8}$.

On the right, Euler dives fearlessly into an intricate series calculation. He writes

$$\frac{1}{\sqrt{1-xx}} = (1-xx)^{-1/2},$$

and applies the generalized binomial theorem to expand the radical as an infinite series. He gets

$$(1-xx)^{-1/2} = 1 + \frac{1}{2}x^2 + \frac{1 \cdot 3}{2 \cdot 4}x^4 + \frac{1 \cdot 3 \cdot 5}{2 \cdot 4 \cdot 6}x^6 + \frac{1 \cdot 3 \cdot 5 \cdot 7}{2 \cdot 4 \cdot 6 \cdot 8}x^8 + \cdots.$$

This is a bit of a tricky step, but it really is the familiar binomial theorem, that

$$(1+a)^n = 1 + \frac{n}{1}a + \frac{n(n-1)}{1 \cdot 2}a^2 + \frac{n(n-1)(n-2)}{1 \cdot 2 \cdot 3}a^3 + \cdots$$

in the case $a = xx$ and $n = \frac{-1}{2}$. If n is a positive integer, then the numerators eventually become zero, and we get a finite sum, but the theorem is still true if n is a fraction. Euler's series converges whenever $|x| < 1$. He integrates and multiplies to get

$$s \, ds = \frac{x \, dx}{\sqrt{1-xx}} + \frac{1}{2 \cdot 3}\frac{x^3 \, dx}{\sqrt{1-xx}} + \frac{1 \cdot 3}{2 \cdot 4 \cdot 5}\frac{x^5 \, dx}{\sqrt{1-xx}} + \frac{1 \cdot 3 \cdot 5}{2 \cdot 4 \cdot 7}\frac{x^7 \, dx}{\sqrt{1-xx}} + \cdots.$$

He knows that if he integrates both sides of this, from $x = 0$ to $x = 1$, he will get $\frac{\pi\pi}{8}$. If he integrates an individual term on the right, using integration by parts, he gets

$$\int \frac{x^{n+2} dx}{\sqrt{1-xx}} = \frac{n+1}{n+2} \int \frac{x^n dx}{\sqrt{1-xx}} - \frac{x^{n+1}}{n+2}\sqrt{1-xx}$$

Since the second term is zero at both endpoints, he can ignore it, and he gets a nice reduction formula. He summarizes the integral result with a list:

$$\int_0^1 \frac{x \, dx}{\sqrt{1-xx}} = 1$$

$$\int_0^1 \frac{x^3 \, dx}{\sqrt{1-xx}} = \frac{2}{3} \int_0^1 \frac{x \, dx}{\sqrt{1-xx}} = \frac{2}{3}$$

$$\int_0^1 \frac{x^5 \, dx}{\sqrt{1-xx}} = \frac{4}{5} \int_0^1 \frac{x^3 \, dx}{\sqrt{1-xx}} = \frac{2 \cdot 4}{3 \cdot 5}$$

$$\int_0^1 \frac{x^7 \, dx}{\sqrt{1-xx}} = \frac{6}{7} \int_0^1 \frac{x^5 \, dx}{\sqrt{1-xx}} = \frac{2 \cdot 4 \cdot 6}{3 \cdot 5 \cdot 7}$$

So, the integral of the expression above (the one that begins $s\,ds$) is

$$\frac{\pi\pi}{8} = \int_0^1 \frac{x\,dx}{\sqrt{1-xx}} + \frac{1}{2\cdot 3}\int_0^1 \frac{x^3\,dx}{\sqrt{1-xx}} + \frac{1\cdot 3}{2\cdot 4\cdot 5}\int_0^1 \frac{x^5\,dx}{\sqrt{1-xx}}$$
$$+ \frac{1\cdot 3\cdot 5}{2\cdot 4\cdot 6\cdot 7}\int_0^1 \frac{x^7\,dx}{\sqrt{1-xx}} + \cdots.$$

Substituting the values for the integrals gives

$$\frac{\pi\pi}{8} = 1 + \frac{1}{3\cdot 3} + \frac{1}{5\cdot 5} + \frac{1}{7\cdot 7} + \frac{1}{9\cdot 9} + \cdots.$$

The series on the right is the sum of the reciprocal of the odd squares, tantalizingly close to the Basel problem, and an easy trick makes it into a solution. Any number is the product of an odd number and a power of 2. For odd numbers, the power of 2 is 2^0. Hence, any square is the product of an odd square and a power of 4. So, Euler multiplies this equation by the sum of the reciprocals of the powers of 4, as follows:

$$\frac{4}{3} = 1 + \frac{1}{4} + \frac{1}{16} + \frac{1}{64} + \frac{1}{256} + \cdots$$

and gets

$$\frac{\pi\pi}{6} = 1 + \frac{1}{4} + \frac{1}{9} + \frac{1}{16} + \frac{1}{25} + \frac{1}{36} + \cdots.$$

It is a different solution to the Basel problem that does not depend on infinite products. In fact, all it requires to meet modern standards of rigor is that we fill in a few routine steps and notice that a few series are absolutely convergent, so that we can do things like multiply two different series together, as we did in the very last step.

References

[E41] Euler, Leonhard, De summis serierum reciprocarum, *Commentarii academiae scientiarum Petropolitanae* 7 (1734/5) 1740, pp. 73–86. Reprinted in *Opera Omnia* Series I vol. 14, pp. 73–86. Available online at `EulerArchive.org`.

[E63] ——, Démonstration de la somme de cette suite $1 + \frac{1}{4} + \frac{1}{9} + \frac{1}{16} + \frac{1}{25} + \frac{1}{36} +$ etc., *Journal littéraire d'Allemagne, de Suisse et du Nord* (La Haye) 2:1, 1743, pp. 115–127. Reprinted in *Bibliotheca Mathematica* 83, 1907–1908 pp. 54–60 and in *Opera Omnia* Series I vol.14, pp. 177–186. Available online at `EulerArchive.org`.

37

Cannonball Curves

(December 2006)

In theory there is no difference between theory and practice. In practice there is.

— Yogi Berra
also attributed to Chuck Reid, Jan L. A. van de Snepscheut, Manfred Eigen, et al.

We all know that the trajectory of a thrown object under the influence of gravity is a parabola, don't we? Isn't that one of the things that got Galileo in so much trouble back in the 1600s? Don't we do dozens of calculus problems every semester based on this fact?

This idea of a parabolic trajectory hasn't always been as obvious and widely accepted as it is today. We will devote this month's column to a brief history of trajectory curves and a bit about Euler's role in their evolution.

Early scholars of ballistics are said to have thought that a cannonball would have a triangular trajectory, though in my brief search of some 16th and 17th century books on gunnery, I could find no primary source that actually made this claim. Still, such a belief in triangular trajectories would be consistent with the scientific views of the time. Remember that we have not always classified chemical elements with the Periodic Table. Once there were only four elements, Earth, Air, Fire and Water. We could explain natural phenomena by the tendency of an object to seek to restore its proper balance of these four elements.

Some phenomena are difficult to explain within the four-element system. For example, if Water is an element, its natural balance ought to be just Water. But then, how can water contain enough of the element Air to evaporate? Fortunately, there were enough taverns that this question could be thoroughly discussed and a satisfactory solution found.

Some phenomena had explanations that seem quite silly today. For example, women were said to be overly prone to crying. This was because they had a tendency to accumulate too much Water, and tears were the natural way for women to re-balance their elements by removing the excess Water.

To the modern reader, such explanations seem strange and unfamiliar, sometimes even incredible, but in their time they were remarkably useful in explaining and even predicting natural phenomena.

In the case of ballistics, here is an argument in favor of a triangular trajectory. Suppose we propel an object like a cannonball into the air. A cannonball is mostly Earth when its elements are correctly balanced. When we propel it into the air, though, we are adding Air to its composition, so it rises. If we could make the Air stay in the cannonball, that would be unnatural, but we could do it with witchcraft. That's one of the things witchcraft could do; force things into an unnatural state, like floating cannonballs. However, our cannonball obeys the laws of nature, and it expels its excess Air. When the Air is all gone, balance is restored, and the cannonball falls straight down, as is its nature.

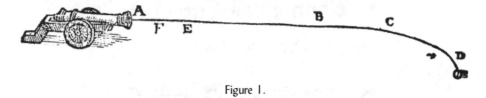

Figure 1.

By the end of the 1500s, Nicolo Tartaglia (famous for his feud with Cardano over the algebraic solution of cubic equations) had come to doubt this triangular trajectory theory. He wrote a book on the theory of ballistics. At the behest of King Henry VIII, that book was promptly translated into English [T] and published in 1588 with a title that began *Three bookes of colloquies concerning the arte of shooting of great and small peeces of artillerie*. Figure 1 shows how Tartaglia thought that a cannonball's trajectory would begin almost straight and would gradually turn downward, rather than making the sudden turn downward described by the triangular theory. Tartaglia's argument is wordy and legalistic. In the style of his times, he thinks that truth will be found by a careful examination of the reasons more than by experimentation or observation. Tartaglia presents reasons in favor of, and opposed to the idea of curved trajectories, almost in the style of a courtroom prosecution and defense, and the reader is expected to act as the judge

One particularly interesting aspect of Tartaglia's argument is that he describes the curve as being "straighter" between A and F than it is between A and C, and that, if points were taken close enough together, the curve could be regarded as straight between those two points. He doesn't actually say it, but the modern reader sees him as coming close to describing a curve as a collection of straight line elements an idea that wasn't actually articulated for at least a hundred years.

Figure 2, also from Tartaglia's *Colloquies*, has the colorful motto *Scientia non habit inimicum prater Ignoratem*, "Science has no enemies except Ignorance." It shows a more elevated trajectory than the one above, and the text tells us that the fireball will fall straight down upon its target.[1] Though the trajectory is clearly a curved one, it cannot be a parabola.

In the early 1600s, Galileo gave us the parabolic trajectory. His story is well known, so we won't repeat it here. I particularly like Berthold Brecht's version.[B] His play, *Galileo*, gives a moving account of the political, scientific and religious issues surrounding Galileo's discoveries, though some think that Brecht is a bit too quick to "bend the truth" to make a good story.

---

[1] One wonders if the victims whose homes are about to be burned think that the science that calculated this trajectory might be their enemy?

Figure 2.

Most of our trajectory problems include a disclaimer like "ignoring air resistance." What happens, though, if air resistance *is* significant, and ought not be neglected? Enter Euler.

Euler wrote a book and three articles on ballistics. Though his contributions are relatively few, they were extremely influential, and they have an interesting story. We begin that story in 1736, when Euler published his two-volume masterwork of physics, the *Mechanica*. Euler worked out in complete detail the mechanics of point masses that Newton had only hinted at. English mathematicians and scientists interpreted Euler's work as criticism of their hero Newton, and some of them responded with bitterness and hostility.

One Englishman, Benjamin Robins, was particularly vitriolic. [R1] A few years later Euler's new employer, Frederick the Great, asked Euler what the best book on mathematical ballistics was. Despite the bad review Robins had given his own book, Euler recommended Robins' book and agreed to translate it from English into German. Euler's "translation" came out in 1745 and is remarkable on a number of levels. First, there is no evidence that Euler knew any English. Second, Robins' book had been 150 pages long. Once Euler got done with adding his comments, it was 720 pages long. In 1777, Hugh Brown translated the book back into English, and in 1783 it was translated into French by someone named Lombard. Napoleon supposedly read that edition and it is one of the things that influenced him to rely so much on his scientists, engineers and mathematicians.

Robins' book of 1742 [R2] contains the passage illustrated below:

(87)

P R O P. VI.

The Track described by the Flight of Shot or Shells is neither a Parabola, nor nearly a Parabola, unless they are projected with small Velocities.

How did Robins know that the trajectory of a cannonball was not a parabola? He had invented a device called a *ballistic pendulum* that measured the speed at which a cannonball left the barrel of the cannon. It was quite a clever and simple device, and my father tells me that when he was younger, every high school physics student in Oklahoma who owned a gun did a laboratory experiment to find the muzzle velocity of his own rifle. Can you imagine doing such an experiment in today's schools?

Robins also wrote:

> FOR we have determined, in the fourth Proposition of the present Chapter, that a Musket-ball ¾ of an Inch in Diameter, fired with half its Weight of Powder from a Piece 45 Inches long, moves with a Velocity of near 1700 Feet in 1″. Now, if this Ball flew in the Curve of a Parabola, its horizontal Range at 45° would be found, by the fifth Postulate, to be about 17 Miles   Now all the practical Writers assure us, that this Range is really short of half a Mile. *Diego Ufano* al-

So much for the parabola. Robins knew that Galileo's claim did not apply to objects moving as fast as cannonballs travel, and he put it in his book. Robins didn't say much about what the trajectory really was, and on this, Euler did not add much in his translation.

Euler revisited the question of trajectories in 1753 in a 40-page article with a title that translates as "Research on the true curve that is described by bodies shot through the air or in any other fluid." [E217] Euler reports that there is theoretical and experimental evidence that air resistance should be proportional to the square of the body's speed through the air. He tells us that Newton had written down the differential equations for such trajectories, but that he had "uselessly tried various ingenious methods to arrive at a solution." He says that his own teacher Johann Bernoulli had been the first to give a solution to the problem.

Euler identifies three forces always acting on a projectile:

1. the accelerating force of gravity, always directed vertically downward

2. the buoyant force of the fluid, always directed upward

3. the resistance of the fluid, always directed against the direction of the motion

It is this last force that Euler assumes is proportional to the square of the velocity. The second force is usually small, but Euler has two reasons to consider it. First, he wants his results to be general enough to describe trajectories through *any* fluid, including water, where buoyancy is significant. Second, he knows from some of his other work that the density of air changes with altitude and weather conditions, and he wants to take that into account.

Euler sets out to find what he can about the nature of the trajectory. He works from Figure 3 below.

Euler divides his discussion into two "branches" of the curve, the *ascending branch*, CNA, and the *descending branch*, AMH. He finds that the x component of the velocity is monotonically decreasing (though he doesn't use those words) and that the descending branch has a vertical asymptote, shown in Figure 3 as EF. This alone would show that

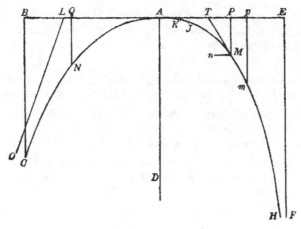

Figure 3.

the curve is not a parabola, but Euler adds two other facts about the descending branch. First, the point at which the curve has its greatest curvature is not at the vertex A, as it would be for a parabola, but is a point on the descending branch near A, labeled K in Figure 3. He also shows that the point where the velocity is minimum is not at A, either, but at a point beyond K, here labeled J.

Next he moves to the ascending branch and shows that when the curve is extended past its initial point at C, it, too, has an asymptote. This asymptote, though, is diagonal, shown in Figure 1 as line LQ. This gives further evidence that the trajectory is not a parabola, for a parabola cannot have a diagonal asymptote either. Also this curve cannot be symmetric, with one vertical asymptote and one diagonal one.

Euler finishes this paper with some tables that show how to find the true curve accurately and how to find various things of interest to artillerymen, like the height of the vertex and the speed of the projectile at various points along its trajectory.

Euler's results are essentially correct, though he does not know about what we now call the Magnus force, and some other forces that turn out to be significant. Let's compare Euler's trajectory to the trajectories that people had proposed earlier.

It's not a parabola. Euler takes pains to show that. Galileo didn't get it.

It's not a triangle. The shape allegedly proposed by the alchemists isn't right either.

It travels diagonally for a while, then curves, and falls almost straight down. Tartaglia, with his ideas of Earth, Air, Fire and Water, came the closest. Of course, it helped that he looked closely at what really happens before he made his predictions.

Let's hear it for four elements.

References

[B]   Brecht, Berthold, *Galileo*, Charles Laughton, tr., Grove Press, NY, 1966.

[E77]   Euler, Leonhard, *Neue Grundsätze der Artillerie aus dem Englischen des Herrn Benjamin Robins übersetzt und mit vielen Anmerkungen versehen*, Berlin, 1745. Reprinted in *Opera Omnia* Series II vol. 14 pp. 3–409. This is one of the few works of Euler that is not yet available through EulerArchive.org.

[E217]   ——, Recherches sur la véritable courbe qui décrivent les corps jettés dans l'air ou dans un autre fluide quelconque, *Mémoires de l'académie des sciences de Berlin*, 9 (1753) 1755, pp. 321–362. Reprinted in *Opera Omnia* Series II vol. 14 pp. 413–447. Available online at `EulerArchive.org`.

[R1]   Robins, Benjamin, *Remarks on Mr. Euler's Treatise of Motion, Dr. Smith's compleat System of Optics, and Dr. Jurin's Essay on Distinct and Indistinct Vision*, London, 1739.

[R2]   ——, *New principles of gunnery: containing the determination of the force of gun-powder, and an investigation of the difference in the resisting power of the air to swift and slow motions*. London, J. Nourse, 1742. Available online through *Early English Books Online*.

[T]   Tartaglia, Nicolas, *Three bookes of colloquies concerning the arte of shooting of great and small peeces of artillerie*, John Harrison, London, 1588. Available online through *Early English Books Online*.

38

Propulsion of Ships

(February 2004)

Long before there were Fields Medals or Nobel Prizes, the great scientific academies of Europe regularly proposed problems, with lucrative cash prizes for the best solutions. There has been an ironic change in the order of events for prizes over the last three centuries. Then, the Academies would meet and decide which questions were important. They would announce the problems, and savants around Europe would make their best efforts. Winners would be announced in a big ceremony, and the losers' entries would be "burned before the assembled Academy."

The Paris Prize was the most coveted of these. In the early and mid 1700s, the Paris problems often involved ships and navigation. For 1727, they posed a problem on the masting of ships, how many masts to use and where in the ship to position the masts. A nineteen year old Euler wrote his essay in 1726, and when the results were published in 1728, he had honorable mention. This sparked Euler's lifelong off-and-on interest in mathematical and physical problems involving ships and navigation. Euler wrote only about a dozen papers on the subject, but he wrote two major books, including his last book, *De la construction et de la manoeuvre des vaiseaux*, published in two volumes in 1773.

An illustration from his paper for the 1727 Paris Prize is shown in Euler's "Figure 5."

We leap ahead to Euler's fifth paper on nautical topics, E137, written in 1748 and published in 1750. Jakob

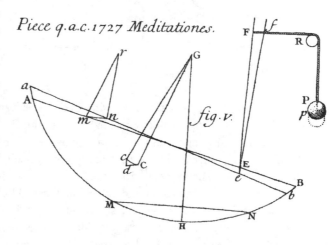

Bernoulli's collected works appeared in 1744, and Euler noticed there an article where Bernoulli proposed a perpetual motion machine to propel a ship. The key idea is illustrated in the figure below.

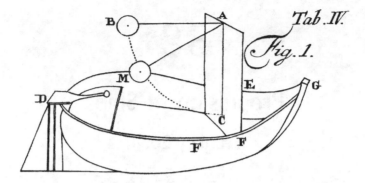

Bernoulli proposed to erect a wall near the bow of the ship and to suspend a heavy weight from the top of the wall at the point marked A. He would lift the ball to the point B and let it drop. It would swing through M, strike the wall at C, and propel the boat forward. At the same time, he hoped, the ball would bounce back to B, from which it would fall again and propel the boat forward again. Obviously, Bernoulli never actually tested his design. It would have been a rough ride.

Euler didn't build the boat, either, but in this article he works through some calculations to show why it couldn't work.

Now let's move on to Euler's seventh nautical paper, E413, written in 1752 for the 1753 Paris Prize, but not published until 1771. It was titled *De promotione navium sine vi venti*, or "On the movement of ships without the force of the wind." Euler's son Johann Albrecht published a French version of this same paper in 1766. In typical essay form, Euler proposes five successively more sophisticated ideas for propelling ships. His first idea, illustrated below, seems almost goofy:

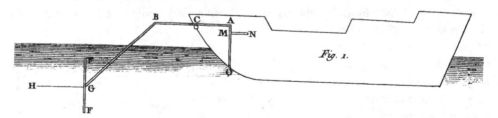

Here, we have a vertical surface, FF, attached to a bent support $ACBG$ and suspended from the front of the boat. Workers inside the boat pull the system along the path MN, then lift it out of the water, push it forward and lower it back in to the water. It works kind of like a hoe. Euler decides it isn't really a very good idea.

Euler's next idea, shown below, isn't very good, either. He proposes attaching paddles, FGF, to both sides of the ship and attaching the paddles to a shaft $DABCCBAD$. Workers on the deck can then turn the shaft and mechanically row the boat. It may seem as if Euler is wasting his time on these silly ideas, but as he studies them, he is developing some equations of energy and fluid resistance that are quite useful.

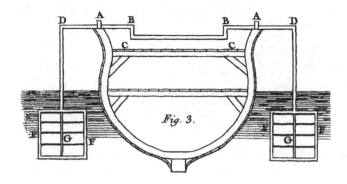

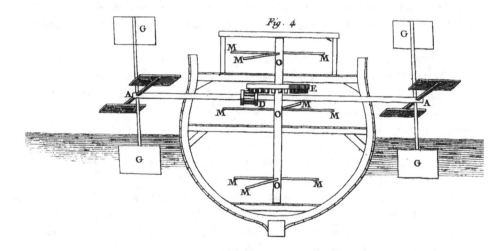

Euler's third idea is a modification of the one-paddle system we just described. He arranges four paddles around the shaft, instead of just one, and he gets a reasonable amount of energy by designing a system of cranks, all labeled *M* in the figure below. This is almost a 19th Century paddle wheel.

Euler's fourth idea also would not prove useful until the 19th Century. Shown below, he proposed extending a kind of fan in front of the ship, then turning it with the same kind of mechanism that he had used for the paddle wheel. This idea was patented in 1838 by the Swedish inventor John Ericsson. Ericsson had in mind that the propeller would be turned by a steam engine.

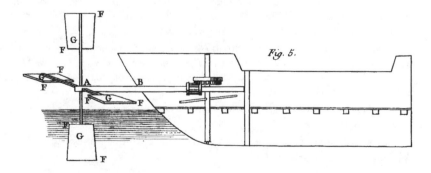

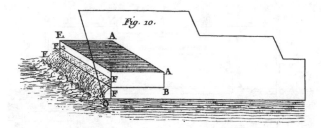

Euler's final idea is almost futuristic. His idea, pictured above, is to try to harness the motion of waves to move a ship. He proposes to put a trough, $AEFB$ just above water level in the stern of the ship.

The idea is that the tops of the waves will be a little higher than the rim of the trough, and the trough will fill with water. Then, since the water in the trough is above the water level around the ship, the water can be drained out of the trough to propel the ship. In principle, the plan should work, but the effect would be very subtle, and there may be problems keeping the ship steady enough to make it work. However, in the 20th Century, electrical power generators were designed and built along the same principles, and they work just fine.

It seems that nobody tried to actually construct a ship that used any of Euler's ideas for propulsion. That is probably just as well, since, as he described them, they almost certainly would not have worked any better than Bernoulli's wall-striking idea would have worked. It is tempting to say, on the basis of these ideas, that Euler deserves credit for "inventing" the paddle wheel boat, the screw propeller and the wave-activated power systems. Not only would that be unfair to Fulton and Ericsson who designed and built steam boats and screw propellers that actually worked, but to credit Euler would be an anachronistic view of history. Euler was using 18th Century ideas to try to solve 18th Century problems. His ideas didn't work. He had no reason to expect that, a hundred years later, steam engines would make ideas similar to his practical, and even important. There is apparently no evidence that Ericsson or Fulton knew of Euler's ideas. They had to rediscover them to solve their 19th Century problems.

References

[E4]   Euler, Leonhard, Meditationes super problemate nautico, *Pièce qui a remporté le I. prix de l'académie royale des sciences en M.DCC.XXVII* (Paris 1728) pp. 1–48. Reprinted in *Opera Omnia* Series II vol. 20, pp. 1–35. Available online at EulerArchive.org.

[E94]   ——, De motu cymbarum remis propulsarum in fluviis, *Commentarii academiae scientiarum Petropolitanae* 10 (1738) 1747, pp. 22–39. Reprinted in *Opera Omnia* Series II vol. 20, pp. 83–100. Available online at EulerArchive.org.

[E413]   ——, De promotione navium sine vi venti (Mémoire sur la manière de suppléer à l'action du vent), *Pièce qui a remporté le II. prix de l'académie royale des sciences en M.DCC.LII* (Paris 1771) pp. 1–47. Reprinted in *Opera Omnia* Series II vol. 20, pp. 190–228. Available online at EulerArchive.org.

[I]   Anon., Islay Wave Power Station, *BBC News*, news.bbc.co.uk/1/hi/sci/tech/1032148.stm.

[E]   Anon., John Ericsson, Inventor, www.pt5dome.com/JohnEricsson.html.

Online links were live on January 18, 2007.

39

How Euler Discovered America

(October 2006)

This month some people celebrate Columbus Day, the anniversary of that day in 1492 when the Italian navigator Christopher Columbus, sailing on behalf of Spain, first spotted the lands of the New World, thinking they were part of China or India. His confusion about where he really was lingers in the name he gave to the inhabitants of the islands he saw, "Indians."

Whether or not Columbus deserves credit for "discovering" America has become a familiar topic of controversy. There are those who point out that Columbus died still thinking that he'd found a shorter route to India and China, not realizing that the Americas were whole continents blocking the way to East Asia. And he probably never set foot on the mainland itself. If he so misunderstood what he had discovered, does he deserve credit for the discovery?

Further, Columbus was probably not the first European to visit America. Norse visitors apparently left remains of their settlements in eastern Canada, a place they called Vinland, and people make similar claims for Irish fishermen, Irish monks, and even for wayward Roman and Egyptian mariners.

The Original Peoples of the Americas, of course, have a considerably different point of view. They had known about the Americas since the beginning of time, so Columbus' discovery was old news to them.

On the other hand, visits by the Norse stopped after a few decades, and other pre-Columbian visitors don't seem to have established a permanent presence. Columbus opened the doors, and Europeans from many countries followed. The Columbus voyages were the ones that changed America. The Norse, the Irish, the Romans, and so forth, had virtually no impact in America or in Europe.

In this column, we will make the (slightly silly) case that Euler deserves some credit for the European discovery of America. The main purpose of the column is to tell that story, but we will also make the small point that this whole "Discovery of America" story can be viewed as a metaphor for mathematical discovery as well.[1]

Let us try to set up some working criteria for a "discovery."[1]

---

[1] Watch closely here, as this is where I'm trying to "pull the wool over your eyes." I will carefully manipulate the definition of "discovery" to carve room to give Euler credit. I do a similar thing when people ask (as they

1. You have to find something you didn't know about before.

2. You have to identify what you have found.

3. You have to announce it, and people have to remember that you found it.

4. You have to demonstrate that what you have found is something new and previously unknown, at least to the people around you.

Credit for a discovery can be shared if people do things together or if some people do some things and other people do others. There is room for debate, discussion and exceptions. These aren't axioms.

By these criteria, Leif Erikson's Norse settlement wasn't a discovery of America because it didn't have a sufficiently lasting effect. People abandoned his settlements and forgot that they had been there.

Columbus deserves only partial credit because he so egregiously misunderstood what he had found. By 1520, though, when Cortez conquered Mexico City, people had figured out that New Spain wasn't India, Japan or China, but a new place altogether. So, it is evident that by 1520, the Spanish had satisfied criteria 1, 2, 3 and part of criterion 4, for a European discovery of America.

What, the reader asks, did the Spaniards miss about Criterion 4? Despite sending expeditions clear around South America and up the Pacific coast to Northern California, they were unable to establish that North America was not attached to Asia. If America were really part of Asia, then India and China could be the Far East, and America would just be Farther East, and not a New World. (Also, Columbus would be right; he'd discovered part of Asia.)

Establishing that America was not Asia fell to the Russians. In the mid-1600s, a Russian Czar sent an expedition to establish the limits of the Russian Empire. They came back a few years later with a report that showed that Siberia ended at the sea, but they did not try to find what, if anything, was beyond that sea. They did, though, establish that Siberia wasn't connected to America. But by the time they returned to Moscow, the Czar who had sent them had died and the new Czar didn't care. They filed the report in the archives, where it rested forgotten for over 200 years.

The issue of the extent of Russia resurfaced shortly after the establishment of the St. Petersburg Academy, Euler's employer from 1728 to 1741 and from 1766 to his death in 1783. Peter the Great, not knowing about the expedition in the 1600s, sent the First Kamchatka Expedition, led by Vitus Bering, to explore the far reaches of Siberia. [Anon1, Anon2] Bering didn't answer all of Czar Peter's questions, though, so Peter's successors sent him to lead the Second Kamchatka Expedition as well, which lasted from 1733 to 1741. Bering himself died on the trip, but survivors reported back in 1743 with, among other things, the news of the Bering Strait that separated Asia from North America

Again, while they were gone, times had changed. Czars and Czarinas had died, and, in the wake of xenophobic riots in St. Petersburg, Euler had left for Berlin and the St. Petersburg Academy had fallen into disarray. Bering's discoveries made Russia the largest country in the world, and at the same time completed the discovery of America by

---

often do) "Who published more, Euler or Erdős?" I ask them who they want to win, and then I devise a scoring method to allow their favorite candidate to win. (Of course, Christian Wolff published more than both of them put together.)

establishing that America wasn't part of Russia. With Russia in chaos, there was nobody in Russia to make the announcement.

Making the announcement fell to Leonhard Euler, still the preeminent member of the St. Petersburg Academy, and really the only member who was still taking his responsibilities seriously. Euler chose to make his announcement in the form of a letter dated December 10, 1746 to his friend Caspar Wetstein, Chaplain and Secretary to the Prince of Wales and member of the Royal Society in London. The following February, Wetstein read the announcement before the Society, and he placed an extract [E107] of the letter in the Society's *Philosophical Transactions* of 1748. The "extract" begins:

XIV. *Extract of a Letter from Mr.* Leonard Euler, *Prof. Mathem. and Member of the* Imperial Society *at* Petersburgh, *to the Rev. Mr.* Cha. Wetstein, *Chaplain and Secretary to His Royal Highness the* Prince *of* Wales, *concerning the* Discoveries *of the* Russians *on the* North-East Coast *of* Asia.

Berlin, Dec. 10. 1746.

Read Feb. 5. —— **A**S you are desirous to hear some-
1746 7.   thing more particular concern-
ing the *Russian* Expeditions to the North and North-East of *Asia*, I will here give you an Account of all that has come to my Knowlege relating to the same.

This is one of Euler's only publications in English. The full text is in the *Opera Omnia*, Series III, volume 2, and an image of the original (from which this image is taken) is available online at The Euler Archive.

Thus, Euler deserves a share of the credit for the discovery of America, for his role in satisfying our Criterion 4 for a discovery.

Is the reader convinced? Must we demand that American history textbooks be rewritten to take out Christopher Columbus and Henry Cabot, to be replaced by Vitus Bering and Leonhard Euler?

Of course not, but it does illustrate how "discovery" is more complicated than we expect.[2] Let us consider mathematical discovery, starting with $V - E + F = 2$.

Some people call the Euler formula, $V - E + F = 2$, relating the vertices, edges and faces of a simple polyhedron, the "Euler-Descartes formula," [S, E230, E231] because Descartes discovered, but did not communicate, a fact about polyhedra that modern anal-

---

[2]Moreover, our choice of the word "discovery" instead of "invention" embeds some assumptions about the nature of mathematics. If we "discover" it, then we are taking a philosophical stance that mathematics exists anyway, whether or not there is a human mind to know it. If we "invent" it, then mathematics is a creation of mankind, and doesn't exist without humans. This is yet another opportunity for debate. Personally, I take both sides at once. I tell people that beautiful mathematics is discovered; ugly mathematics is invented. Euler discovered.

ysis can show to imply the Euler formula. Euler's contributions fully meet the criteria for a discovery. Descartes' falls far short.

On the other hand, in 1777 Euler wrote a paper [E703] in which he expanded a formula describing planetary orbits, $b/(1 + e \cos \varphi)$, as a cosine series that he wrote as

$$\Gamma : \varphi = A + B \cos \varphi + C \cos 2\varphi + \text{etc.}$$

This is clearly a Fourier series and the coefficients A, B, C, etc. are Fourier coefficients.

Still, Euler thought that his result was one about planetary orbits, and not about periodic functions. That was left to Fourier, whose name, indeed, belongs on the series.

And so it goes; there are a great many more examples in the history of mathematics in general and the discoveries of Euler in particular. They show that most things in mathematics are correctly, or at least reasonably named, and the exceptions make good material for interesting talks and articles on the history of mathematics.

References

[Anon1] Anonymous, Second Kamchatka expedition, *Wikipedia*, en.wikipedia.org/wiki/ Second_Kamchatka_expedition, consulted September 22, 2006.

[Anon2] Anonymous, Vitus Bering, *Wikipedia*, en.wikipedia.org/wiki/Vitus_Bering, consulted September 22, 2006.

[E107] Euler, Leonhard, Extract of a letter from Mr. Leonhard Euler, Prof. Mathem. and Member of the Imperial Society at Petersburgh, to the Rev. Mr. Cha. Wetstein, Chaplain and Secretary to His Royal Highness the Prince of Wales, concerning the Discoveries of the Russians on the North-East Coast of Asia, *Philosophical Transactions* (London) 44, 1748, pp. 421–423, reprinted in *Opera Omnia* Series III vol. 2 pp. 373–375. Available online at EulerArchive.org.

[E230] ——, Elementa doctrinae solidorum, *Novi commentarii academiae scientiarum Petropolitanae* 4 (1752/3) 1758, pp. 109–140, reprinted in *Opera Omnia* Series I vol. 26 pp. 71–93.

[E231] ——, Demonstratio nonnullarum insignium proprietatum quibus solida hedris planis inclusa sunt praedita, *Novi commentarii academiae scientiarum Petropolitanae* 4 (1752/3) 1758, pp. 140–160, reprinted in *Opera Omnia* Series I vol. 26 pp. 94–108.

[E703] ——, Methodus facilis inveniendi series per sinus cosinusve angulorum multiplorum procedentes, quarum usus in universa theoria astronomiae est amplissimus, *Nova acta academiae scientiarum Petropolitanae*, 11 (1793), 1798, pp. 94–113. Reprinted in *Opera Omnia* Series I vol. 16, pp. 311–332. Available online at EulerArchive.org.

[S] Sandifer, Ed, *V*, *E* and *F*, part 1, *How Euler Did It*, MAA Online, www.maa.org, June 2004. Reprinted in this volume on pp. 9–12.

40

The Euler Society

(May 2005)

Every summer since 2002, twenty or so Euler fans have gathered for the three-day annual meeting of The Euler Society. The first meeting was in Rumford, Maine, and subsequent meetings have been at the Conference Center at Roger Williams University in Portsmouth, RI. This year's meeting will return to Roger Williams from the evening of Sunday, August 7 until lunchtime the following Wednesday. This should be timed so that people have about 30 hours to get to Rhode Island from the MathFest in Albuquerque.[1]

The featured talk of the conference is always The Euler Lecture. The 2004 Euler Lecture was by Rüdiger Thiele, who teaches in Leipzig and lives 20 km to the west in Halle, just two blocks from where Cantor lived. Euler has a connection to Halle. His son attended university there and Euler himself escorted his son when he went away to college, along some of the very roads Rüdiger drives to work.

Rüdiger's talk, *Leonhard Euler: 1750–1760*, dwelt mostly on Euler's life and times in the 1750s. His life was strongly influenced by three main factors, the European Enlightenment, his own interests in mathematics and natural sciences, and his own deep sense of Christian tradition. This particular decade had an additional, and mostly unmentioned influence: Euler's mother lived in his house near Berlin from 1750 to 1761.

Euler had immense responsibilities in the Berlin Academy. He supervised the observatory, the library, preparation of almanacs, publications and the botanical gardens. The preparation of almanacs was particularly sensitive, since the sale of those almanacs was the principal source of revenue for the entire Academy. He also advised King Frederick on diverse technical matters regarding mines, lotteries, canals and fountains, all while essentially running the St. Petersburg Academy by mail, writing endlessly, and serving a key role in his Parish Council.

Despite all this service, Euler was unpopular in Frederick's court, in part because courtiers valued wit over wisdom, and in part because Frederick himself viewed science as a servant of the ends of the State.

---

[1] Of course this is now old news. The College of St. Rose in Albany NY hosted the 2006 meeting. As I revise these columns for publication, The Euler Society is planning its 2007 meeting as a joint meeting with the MAA at the MathFest in San Jose, CA.

Three days after The Euler Society Conference, at the MathFest in Providence, the MAA presented Rüdiger with its Lester R. Ford award for his article "Hilbert's Twenty-Fourth Problem," which appeared in the *Monthly* in January 2003.

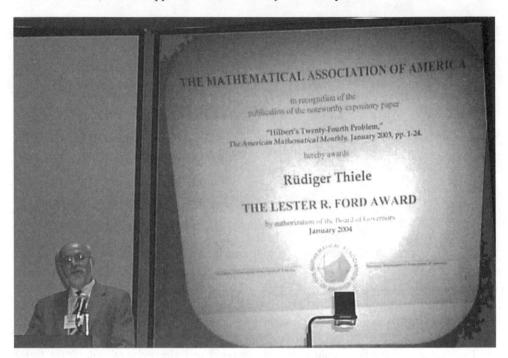

Larry D'Antonio of Ramapo University in New Jersey, gave the first talk of The Euler Society Conference, *Euler's Christmas Present of 1751*. Larry has a variety of interests, from bioinformatics to some topics in medieval Arab mathematics, as well as being the Coordinator for Student Chapters for the New Jersey Section of the MAA. This day, though, Larry was telling the story of the *Collected Works* of Count Giulio Carlo de Toschi de Fagnano. Fagnano (1682–1766) did his best work in elliptic integrals, though his name is also attached to "Fagnano's identity,"

$$\ln \sqrt{\frac{1-i}{1+i}} = -\frac{\pi i}{4}.$$

Two days before Christmas 1751 (a holiday Euler would have observed with more piety than celebration), Euler received a copy of Fagnano's *Collected Works*, containing, in a very modern "theorem-proof" style, a great many new results in elliptic integrals, including the so-called addition formula for elliptic arc lengths and the "trisection" of the lemniscate. Fagnano's ideas so inspired Euler that he spent the next month preparing his article [E252] "Observations on the comparison of arcs of irrectifiable curves." He presented the article to the Academy on January 27, just five weeks after he received Fagnano's *Works*, though it wasn't published until 1761. In this paper, and enough other papers to fill two full volumes of the *Opera Omnia*, Euler presented and extended Fagnano's results and applied them to diverse subjects as the oscillation of pendulums with large amplitudes, the arithmetic of elliptic integrals, the measurement of the earth, and the three-body problem.

John Glaus of Rumford, Maine, described the long and fascinating correspondence between Euler and Johann Kaspar Wettstein (1695–1759), Chaplain to the Prince of Wales and later to his widow. Wettstein helped distribute the Berlin Academy's almanacs in England. In these letters, Euler describes his wonderful relationship with King Frederick when he first arrived in Berlin, and we can see the gradual deterioration of their respect for each other. Euler also gives some details not found elsewhere of the role he played in the expedition of Vitus Bering that showed that Alaska was not connected to Asia and that California was not an island.

Karin Reich of the University of Hamburg spoke on Euler and Gauss. She studied Gauss's books and the circulation records of the library at Göttingen to document the great extent to which Gauss studied Euler. She told an anecdote from 1798, when the 21-year old student Gauss had a bit of money and, like Erasmus[2], spent it to buy four books by Euler. In one of those books, Euler's great 1744 treatise on the calculus of variations, the *Methodus inveniendi*, immediately after the so-called *Theoria elegantissimum*, there is a careful tracing of a portrait of Euler. Karin speculates that Gauss himself traced it, though maybe it was by one of his friends. Gauss often wrote and drew in his books.

Stacy Langton from the University of San Diego spoke on Jakob Bernoulli's seminal probability book *Ars Conjectandi*. Euler Society President Rob Bradley of Adelphi spoke on *Euler, D'Alembert and the Vibrating String Problem*. Lee Stemkoski of Dartmouth talked about recent additions to The Euler Archive, and Vicki Hill showed her wonderful documentary film on Constantin Carathéodory. Carathéodory, in addition to being a fascinating individual and a towering 20th century mathematician, also edited the *Opera Omnia* volumes on the calculus of variations, Series I volumes 24 and 25. Your columnist suggested 23 problems for further research on Euler.

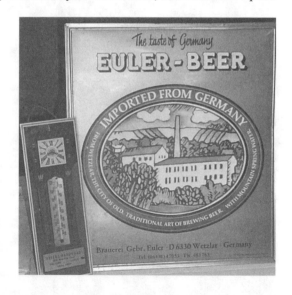

Among the unique aspects of The Euler Society Conferences are the Original Sources Workshops. As a group, those who are interested (and that's pretty much everyone) assemble after dinner with a refreshment and a copy of an original source from Euler's life or work. We read it over. Those with the necessary language skills read in the original language and translate for us, and we discuss what we've read. Usually it is not what we expected to find.

Last year we read Goldbach's famous letter to Euler in which he first makes Goldbach's Conjecture, that every even number, starting with 6, is the sum of two odd primes. We were delighted to read that that's not exactly how Goldbach phrased it.

---

[2]Erasmus is often quoted as saying "If I get some money, I spend it on books. If there is any left over, I buy food, clothing and shelter." Whether or not the quote is accurate, it is a popular sentiment.

A very smart man from Berlin
Said, "Constructions and curves
                    are akin;
But some, without doubt,
Leave parts of curves out,
While others put extra curves in!"

We also looked at an as-yet unpublished letter of Euler to Cramer, part of the conversation on Cramer's Paradox[3] that formed the basis of this column last August. That column was based, in part, on this letter. That workshop apparently inspired the anonymous limerick above that appeared the next morning before breakfast.

Perhaps some readers will want to join this summer's meeting. It is a congenial group, a mixture of people who specialize in Euler and people who are just interested. Details are available (or will be soon) at the Society's website, EulerSociety.org. Maybe you'll be among the happy faces in this year's group picture.

Back row: Lee Stemkoski, Larry D'Antonio, Ed Sandifer, Vikki Hill, Rüdiger Thiele, Karin Reich. Middle row: Bob Kelly, Bruce Burdick, Stacy Langton, Maryann McLoughlin, Kim Plofker. Front row: Rob Bradley, John Glaus.

---

[3] Cramer's Paradox, *How Euler Did It*, August 2004. Reprinted in this volume on pp. 37–42.

References

[OC] O'Connor, J. J., and E. F. Robertson, Guilio Carlo Fagnano dei Toschi, The Mac-Tutor History of Mathematics archive, www-gap.dcs.st-and.ac.uk/~history/ Mathematicians/Fagnano_Giulio.html.

[E252] Euler, Leonhard, Observationes de comparatione arcuum curvarum irrectificabilium, *Novi commentarii academiae scientiarum Petropolitanae* 6 (1756/57) 1761, pp. 58–84, reprinted in *Opera Omnia* Series I vol. 20 pp. 80–107. Available online at EulerArchive.org.

Index

About the Author

Ed Sandifer is Professor of Mathematics at Western Connecticut State University. He earned his PhD at the University of Massachusetts under John Fogarty, studying ring theory. He became interested in Euler while attending the Institute for the History of Mathematics and Its Uses in Teaching, IHMT, several summers in Washington DC, under the tutelage of Fred Rickey, Victor Katz and Ron Calinger. Because of a series of advising mistakes, as an undergraduate he studied more foreign languages than he had to, and so now he can read the works of Euler in their original Latin, French and German. Occasionally he reads Spanish colonial mathematics in its original as well. Now he is secretary of The Euler Society, and he writes a monthly on-line column, *How Euler Did It*, for the MAA. He has also written *The Early Mathematics of Leonhard Euler*, also published by the MAA, and edited, along with Robert E. Bradley, *Leonhard Euler: Life, Work and Legacy*. He and his wife, Theresa, live in a small town in western Connecticut, and he has run the Boston Marathon every year since 1973.